Macintosh Version

Interactive Mechanics

An Interactive Physics Workbook

Alejandro L. Garcia

San Jose State University

PRENTICE HALL Upper Saddle River, NJ 07458

Production Editor: ***James Buckley***
Production Supervisor: ***Joan Eurell***
Acquisitions Editor: ***Alison Reeves***
Production Coordinator: ***Donna Sullivan***
Art Director: ***Heather Scott***
Cover Design: ***Paul Gourhan***

Simon & Schuster/A Viacom Company
Upper Saddle River, NJ 07458

Printed in the United States of America

10 9 8 7 6 5 4 3 2 1

ISBN 0-13-261710-2

Prentice-Hall International (UK) Limited, *London*
Prentice-Hall of Australia Pty. Limited, *Sydney*
Prentice-Hall Canada, Inc., *Toronto*
Prentice-Hall Hispanoamericana, S.A., *Mexico*
Prentice-Hall of India Private Limited, *New Delhi*
Prentice-Hall of Japan, Inc., *Tokyo*
Simon & Schuster Asia Pte. Ltd., *Singapore*
Editora Prentice-Hall do Brasil, Ltda., *Rio de Janeiro*

Table of Contents

Project 1	Units and Measurement	1
Project 2	Simple Motion in One Dimension	7
Project 3	Free Fall as One Dimensional Motion	13
Project 4	Vectors	19
Project 5	Motion on a Curved Path	27
Project 6	Velocity and Acceleration I	35
Project 7	Velocity and Acceleration II	41
Project 8	Projectile Motion	49
Project 9	Forces in One Dimensional Motion	57
Project 10	Forces in Two Dimensional Motion	65
Project 11	Centripetal Force	73
Project 12	Work and Kinetic Energy	81
Project 13	Conservation of Energy	87
Project 14	Momentum and Center of Mass	93
Project 15	Probability and Statistics	99

Preface for Students

Mechanics is the study of how and why objects move. While the ancient Greeks mastered much of geometry and algebra, understanding something as simple as how an apple falls to Earth came 15 centuries later. Physics advanced quickly only after the introduction of the scientific method of observation and measurement.

This workbook is designed to help you understand basic mechanics through the observations and measurement of simple phenomena (e.g., a block sliding down an incline, a ball flying through the air). The emphasis is on concepts rather than calculation. I find that my students have few problems with the computations in a homework exercise once they understand the fundamental ideas (the "getting started" phase).

Ideally, you would like to observe physical phenomena in their real-world setting, yet this is often difficult and clumsy to do. If you drop a pencil, how far does it fall in the first tenth of a second? What is its speed? To overcome these difficulties, this workbook is designed to use computer simulations. You have great control in the simulation world: slow down time or even reverse it, turn gravity on and off, run experiments over and over with the press of a button.

Useful as they are, you will learn little from passively watching the simulations; you need to work through the questions and problems in the workbook projects. These projects cover the basic topics of elementary mechanics. Once you have mastered mechanics, you have the foundation upon which the rest of physics is built.

Preface for Instructors

The projects in this workbook are similar to the mechanics experiments performed in an introductory physics course, except that *Interactive Physics®* computer simulations replace laboratory equipment. The level is suitable for high school or university physics; at San Jose State University, we use the workbook in our physics prep course. The projects are designed to be self-contained and can be assigned as additional homework. Alternatively, the workbook can be integrated into a lab section to augment or replace hands-on experiments. A student should be able to complete each project in about one and a half hours with little or no supervision. The computer resources required are modest (a Macintosh® 68030 computer or a 486 PC running Windows®) and students may use their home computers.

IMPORTANT: A COPY OF THE INTERACTIVE PHYSICS PROGRAM IS REQUIRED TO RUN THE SIMULATIONS IN THIS WORKBOOK.

Name: ______________________________ Instructor: __________________

Partner: ______________________________ Section Day/Time: ____________

Project 1
Units and Measurement

Objectives

In this first project, you are asked to make a few simple observations and perform some basic calculations. The topics covered in this project include:

- Running *Interactive Physics®*
- Units and unit conversion
- Scientific notation

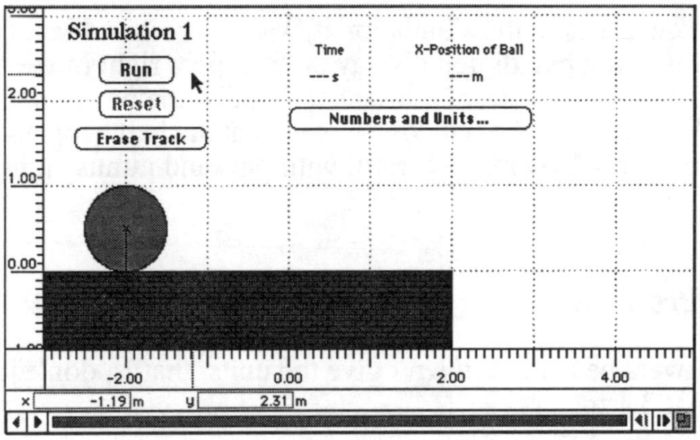

Procedure

Using the *Interactive Physics* program, run the simulation for this project (with the mouse, click on the **Run** button). You observe a steel ball rolling off of a platform. The first time you run the simulation, the program runs slowly since it has to calculate the motion of the objects. If you reset (press the **Reset** button) and run the simulation again, you will notice that it runs faster and more smoothly the second time. This is because the motion has been "filmed" and you are viewing the recorded film.

To stop the simulation, press the **Run** button (which becomes a **Stop** button when the simulation is running) or click the mouse in any blank region (notice that the cursor turns into a stop sign). If the simulation time exceeds 3 seconds, it stops automatically.

After the simulation stops, you may use the VCR-like controls at the bottom of the window to rewind, play, or step through individual frames.

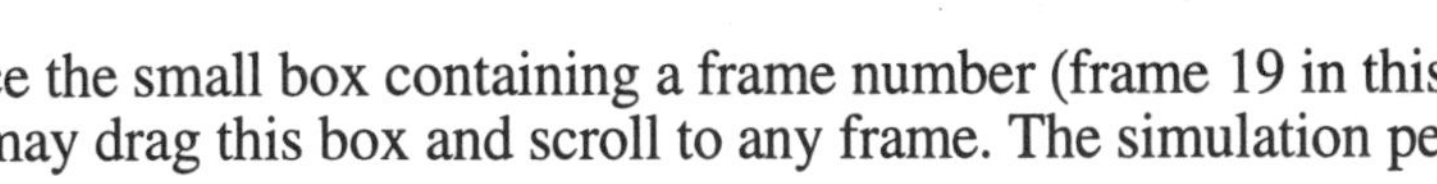

Notice the small box containing a frame number (frame 19 in this case); using the mouse you may drag this box and scroll to any frame. The simulation periodically traces the outline of the ball to help you understand its motion (after resetting the simulation, hit the **Erase Trace** button to remove the outline traces).

By default, the simulation uses the metric system (also known as SI) for measurements. Press the **Numbers and Units** button to change units; click on the menu list under **Unit System:** to see a list of other systems.

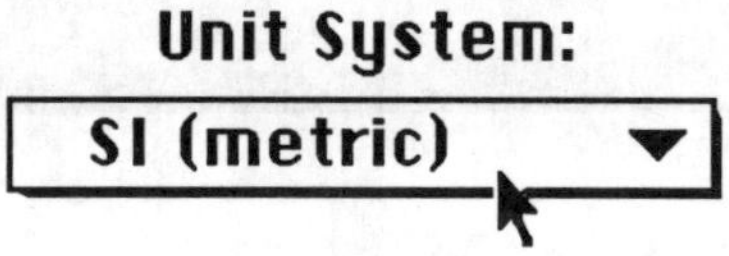

For example, select **English (Slugs)** and distances are measured in feet instead of meters. Note: for SI units, the menu may list **SI (radians)** or **SI (metric)**.

IMPORTANT: ALWAYS **RESET** THE SIMULATION BEFORE CHANGING UNITS (OR MAKING ANY OTHER CHANGES).

You can determine distances in the simulation by various means. First, the horizontal position of the ball is given in a digital display on the upper right of the screen. Second, you may use the rulers on the sides of the screen. Finally, the position of the mouse pointer is displayed near the bottom of the window. Notice that the ball happens to start at position $x = -2.00$ m and *not* at $x = 0.00$ m; be careful with plus and minus signs when measuring distances.

Questions and Problems

NOTE: In your answers be sure to *always* give the units, that is, don't just answer 3.1 if you mean 3.1 feet or 3.1 meters.

1. What is the x coordinate of the ball (in meters) at time $t = 1.00$ second?

2. From your answer to questions #1, use the conversion 1 ft = 0.3048 m to convert the x coordinate of the ball at time $t = 1.00$ s into feet.

3. Reset the simulation. Now use the **Numbers and Units** button to change the measurement system to **English (Slugs)**. Run the simulation again; what value does it give for the x coordinate of the ball (in feet) at time $t = 1.00$ second? (Check with your answer to the previous question).

4. Using the mouse to help you make measurements, find diameter of the ball in feet.

5. What is the radius of the ball in feet?

6. What is the volume of the ball in cubic feet? The volume of a sphere is $V = {}^4/_3\, \pi\, r^3$ where r is the radius.

7. Derive the equation that gives the radius of a sphere in terms of its volume, that is, if $V = {}^4/_3\, \pi\, r^3$, then r = ???.

8. Consider the following units conversion:

1 square foot = $(0.3048 \text{ meters})^2$ = 0.09290 square meters

or

$1 \text{ ft}^2 = 0.09290 \text{ m}^2$

Following this example, determine the number of cubic meters in one cubic foot, that is, 1 ft^3 = ??? m^3.

9. Using your results from questions #6 and #8, determine the volume of the ball in cubic meters.

10. From the equation you derived in question #7 and your answer to question #9, determine the radius of the ball in meters.

11. Convert your answer in question #10 to feet. Does your answer agree with question #5? Could round-off error account for any discrepancy?

12. Reset the simulation. Now use the **Numbers and Units** button to change the measurement system to **Atomic**. The unit of length is now the nanometer or nm (e.g., the diameter of an oxygen molecule is about 0.4 nm). From your textbook, how many meters in a nanometer?

13. The digital displays now use scientific notation but in a slightly different form than is used in textbooks. The examples below illustrate the two styles:

Textbooks:	3.0×10^8	6.2×10^{-21}
Computer:	3.0e+8	6.2e-21

For what reasons might the digital display on a computer use a different notation?

14. Run the simulation. What is the x coordinate of the ball (in nm) at time $t = 0.50$ seconds? At time $t = 0.80$ s?

15. From your answers in question #14, what is the total distance traveled by the ball during those 0.3 seconds?

16. The average speed of an object is defined as

$$\text{Average speed} = \frac{\text{Distance traveled}}{\text{Time elapsed}}$$

From your answer in question #15, what is the average speed (in nm/s) of the ball as it rolls on the floor during those 0.3 seconds?

17. Convert the speed computed in question #16 from nm/s to m/s.

18. Just looking at the traces of the ball's motion, I would say that it rolls on the floor with a constant speed. What visual clues tell me that the speed is constant?

19. If the ball rolls with constant speed (computed in questions #16 and #17), how many meters does it travel between $t = 0.00$ s and $t = 1.50$ s?

20. Reset the simulation. Use the **Numbers and Units** button to change the measurement system back to **SI (metric)** units. Run the simulation and check your answer to the previous question. From the values given by the simulation, how far does the ball travel horizontally between $t = 1.50$ s and $t = 3.00$ s?

Name: ______________________________ Instructor: ____________________

Partner: ______________________________ Section Day/Time: ______________

Project 2
Simple Motion in One Dimension

Objectives

In this project, you observe the motion of an object in one dimension. The topics covered in this project include:

- Velocity
- Acceleration
- Graphing

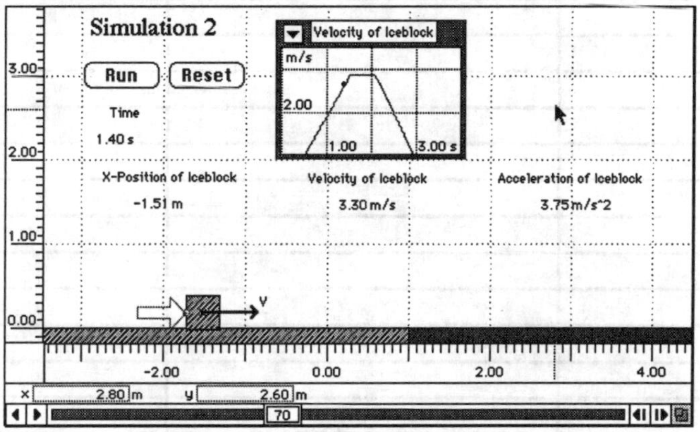

Procedure

Using the *Interactive Physics* program, run the simulation for this project. The first run is slow since the motion must be calculated from scratch. Reset the simulation (press the **Reset** button) and run it again, carefully observing the motion.

In the simulation, a large block of ice sits on the smooth, frictionless surface of a frozen pond. Initially, the block is at rest (i.e., it is not moving). At time $t = 0.5$ s, a constant force begins pushing on the block, giving the block a constant acceleration. The force is indicated by the large arrow to the left of the block. A thin arrow indicates the velocity of the block; notice how this arrow lengthens as the velocity increases.

At time $t = 1.5$ s, the force stops pushing and the block slides freely with a constant velocity.* When the block reaches the shore, friction begins acting† and the block's velocity begins decreasing. Eventually, the block comes to rest and the arrow indicating its velocity shrinks to a dot.

The digital meters measure the time and the block's position, velocity and acceleration. The velocity is also graphed as a function of time.

---

* If you look carefully, you will still see a faint image of the force arrow when the force is turned off.

† Note that the simulation does *not* draw an arrow to indicate the frictional force acting on the block.

Questions and Problems

1. Run the simulation and fill out Table I. You will probably want to use the VCR-like controls (located at the bottom of the window) to step through the appropriate frames.

2. Using the data from Table I, plot the velocity of the block versus time (i.e., plot the first and second columns of the table above) on Graph 1 (see page 11). Your graph should be similar to the graph made by the simulation. Be sure to graph *all* your data.

3. Using the data from Table I, plot the acceleration of the block versus time (i.e., plot the first and third columns of the table below) on Graph 2 (see page 12). Be sure to graph *all* your data.

Table I: Velocity and acceleration of ice block at various times.

Time (seconds)	Velocity (m/s)	Acceleration (m/s^2)
0.00		
0.20		
0.40		
0.60		
0.80		
1.00		
1.20		
1.40		
1.60		
1.80		
2.00		
2.20		
2.40		
2.60		
2.80		
3.00		
3.20		
3.40		
Graphs 1 and 2 Horizontal axis	Graph 1 Vertical axis	Graph 2 Vertical axis

4. In Table I, during which periods of time is the block at rest? Give numerical values such as "between time $t = 0.0$ seconds and ..."

5. In Table I, during which periods of time does the block move with constant velocity (careful, an object at rest has a constant velocity of zero)?

6. In Table I, during which periods of time is the acceleration equal to zero?

7. From the simulation, when is the position of the center of the block equal to zero?

8. What is the velocity of the block at time $t_1 = 0.7$ s? We will call this velocity v_1.

9. What is the velocity of the block at time $t_2 = 1.4$ s? We will call this velocity v_2.

10. The average acceleration of an object is defined as

$$\text{Average acceleration} = \frac{\text{Final velocity} - \text{Initial velocity}}{\text{Final time} - \text{Initial time}} = \frac{v_2 - v_1}{t_2 - t_1}$$

Using your results from questions #8 and #9, compute the average acceleration of the block during that period of time. Compare your answer to the values of instantaneous acceleration given by the simulation.

11. Repeat questions #8 and #9 but for $t_1 = 1.7$ s and $t_2 = 2.0$ s.

12. Repeat question #10 but for $t_1 = 1.7$ s and $t_2 = 2.0$ s.

13. Repeat questions #8 and #9 but for $t_1 = 2.2$ s and $t_2 = 2.7$ s.

14. Repeat question #10 but for $t_1 = 2.2$ s and $t_2 = 2.7$ s.

15. In this simulation, the acceleration happened to be positive when the block's speed was increasing and negative when its speed was decreasing. Explain why this relation is *not* true in general (i.e., an object can have a negative acceleration while its speed is increasing). Hint: Speed and velocity are related but not equivalent; see your textbook. Another hint: Consider the speed, velocity and acceleration of a falling body.

Graph 1: Velocity of the block versus time (question #2)

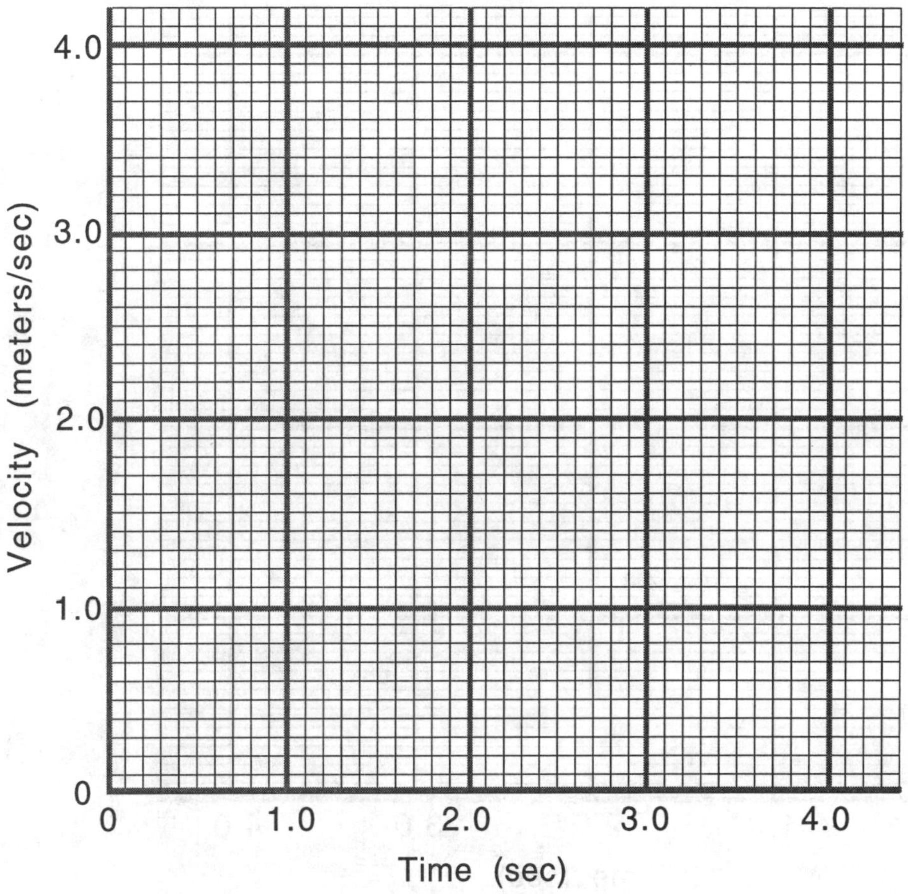

- Graph each data point as a distinct, small dot.
- To increase the visibility of the data, circle each dot, as shown below

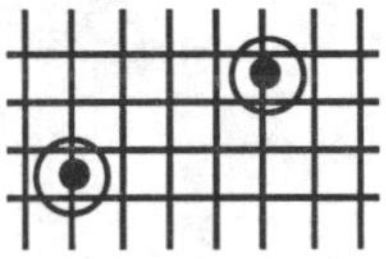

- For this graph, do *not* connect the data points with lines.
- Be sure to graph all of your data.

Graph 2: Acceleration of the block versus time (question #3)

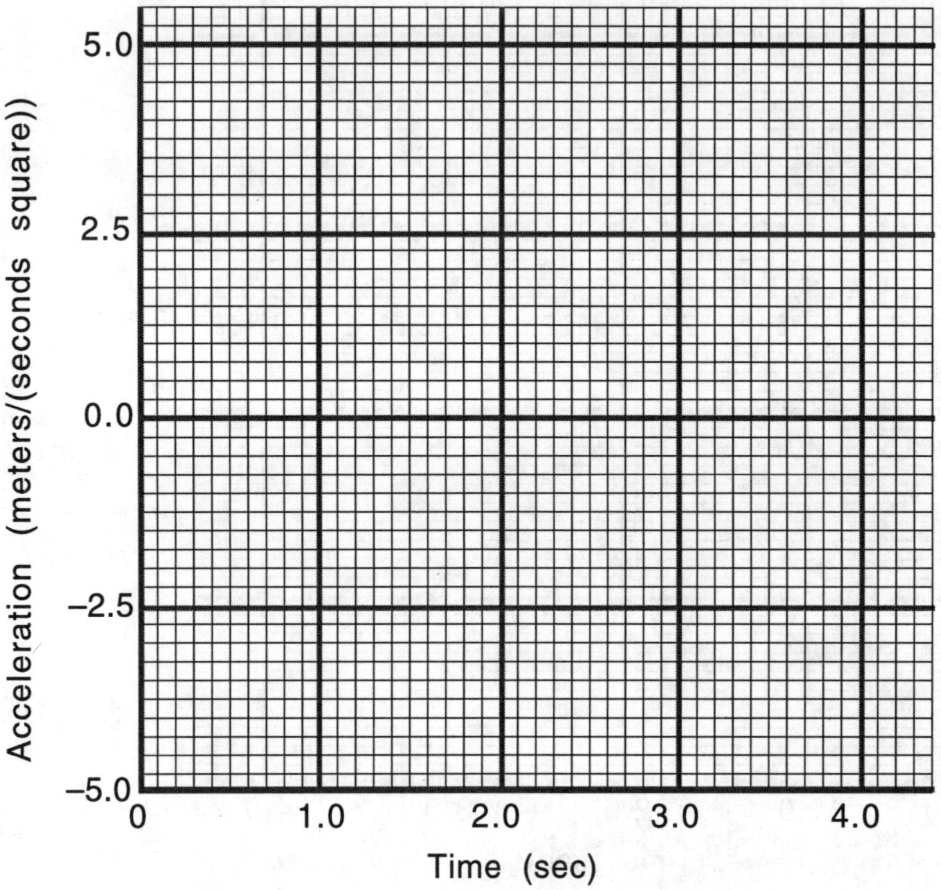

- Graph each data point as a distinct, small dot.
- To increase the visibility of the data, circle each dot, as shown below

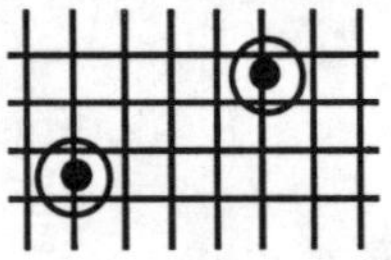

- For this graph, do *not* connect the data points with lines.
- Be sure to graph all of your data.

Name: ______________________________ Instructor: ____________________

Partner: ______________________________ Section Day/Time: ____________

Project 3
Free Fall as One Dimensional Motion

Objectives

In this project, you observe the motion of an object in one dimension. The topics covered in this project include:

- Velocity
- Acceleration
- Slope of a line

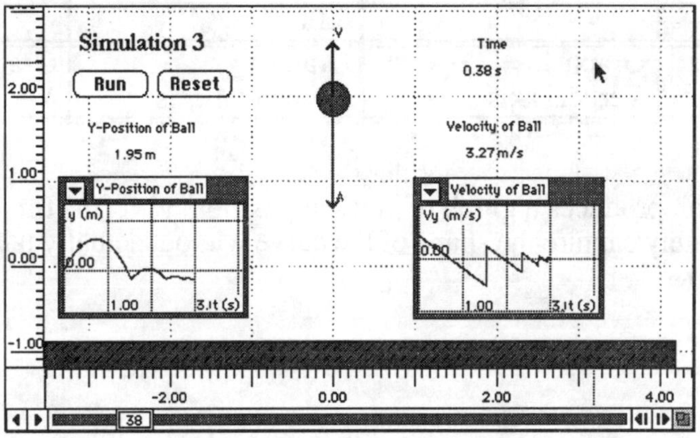

Procedure

Using the *Interactive Physics* program, run the simulation for this project. A rubber ball is thrown upward; the ball falls back downward and bounces on a wooden floor. Run the simulation several times and carefully observe the motion.

Notice that the velocity and acceleration of the ball are marked by arrows (or *vectors*). Initially, the ball's velocity is upward. As the ball rises to its maximum height, the velocity decreases to zero. When the ball begins to fall, the velocity becomes negative and its magnitude increases with time. The force of impact reverses the ball's direction of motion when it strikes the floor. The ball moves upward again but does not reach the same height since some of the ball's energy is lost in the impact.

Gravity is always acting on the ball giving it a constant downward (negative) acceleration. When the ball hits the floor, a very brief but very strong force is exerted on the ball. This force causes the ball's velocity to change suddenly, that is, it causes a large acceleration.

There are two digital meters indicating the ball's position and velocity (in the vertical direction). Below the digital meters, these two quantities are also plotted as a function of time.

Questions and Problems

1. Run the simulation and fill out Table I. You will probably want to use the VCR-like controls to step through the appropriate frames.

2. Using the data from Table I, plot the velocity of the ball versus time on Graph 1 (see page 18). Your graph should be similar to the graph made by the simulation.

Table I: Position and velocity of the ball at various times.

Time (seconds)	Velocity (m/s)	Time (seconds)	Velocity (m/s)
0.00		1.60	
0.20		1.80	
0.40		2.00	
0.60		2.20	
0.80		2.40	
1.00		2.60	
1.20		2.80	
1.40		3.00	
Graph 1 Horizontal axis	Graph 1 Vertical axis	Graph 1 Horizontal axis	Graph 1 Vertical axis

3. The simulation produces a graph of vertical position, *y*, versus time *t*. Sketch this graph; try to accurately capture the shape of the curve without actually taking data.

4. At about what time does the ball reach its maximum height?

5. The ball goes up and then comes back down. At about what time does the ball return to its initial position (at $y = 0.00$ m)?

6. How are the answers to questions #4 and #5 related? Give a simple justification for your answer.

7. What is the ball's velocity when the ball reaches its maximum height?

8. When the ball reaches its maximum height, the velocity vector shrinks to zero length (it becomes a point). Does the acceleration vector also have zero length; if not, in what direction does it point?

9. Does the ball have a positive, negative or zero acceleration when it reaches its maximum height?

10. What is the ball's velocity when it returns to its initial position (at $y = 0.00$ m)?

11. How are the ball's initial velocity and your answer to question #10 related? Give a simple justification for your answer.

12. The average acceleration of an object is defined as

$$\text{Average acceleration} = \frac{\text{Second velocity - First velocity}}{\text{Second time - First time}}$$

or

$$\bar{a} = \frac{v_2 - v_1}{t_2 - t_1}$$

Using t_1= 0.10 s as the first time and t_2 = 0.50 s as the second time, find the velocity of the ball at these two times and use the above equation to compute the ball's average acceleration during that period of time.

13. Repeat the calculation in question #12 but using t_1 = 0.10 s and t_2 = 1.10 s.

14. On your graph of velocity versus time (Graph 1), draw a right triangle using the points at t_1= 0.10 s and t_2 = 0.50 s as the hypotenuse (see sketch below). Explain why the legs of this triangle are the numerator and denominator for the equation in question #12.

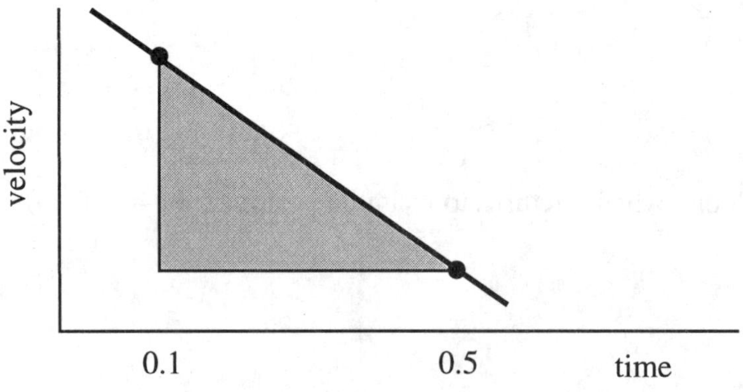

15. On your graph of velocity versus time (Graph 1), draw another right triangle using the points at $t_1 = 0.10$ s and $t_2 = 1.10$ s as the hypotenuse. Explain why the accelerations computed in questions #12 and #13 are equal in value given the form of the triangles in the two cases. The two triangles are not identical yet what simple geometric properties do they have in common?

16. The equation of a line in the *x-y* plane may be written as $y = m\,x + b$, where m is the slope and b is the intercept. Given two points on the line, (x_1, y_1) and (x_2, y_2), find an equation that gives the slope m. Hint: Since the points are on the line $y_1 = m\,x_1 + b$ and $y_2 = m\,x_2 + b$; combine these two equations and solve for m.

17. From your result for the previous question, explain why the average acceleration is the slope of the line on the velocity versus time graph (see question #12).

Graph 1: Velocity of the ball versus time (question #2)

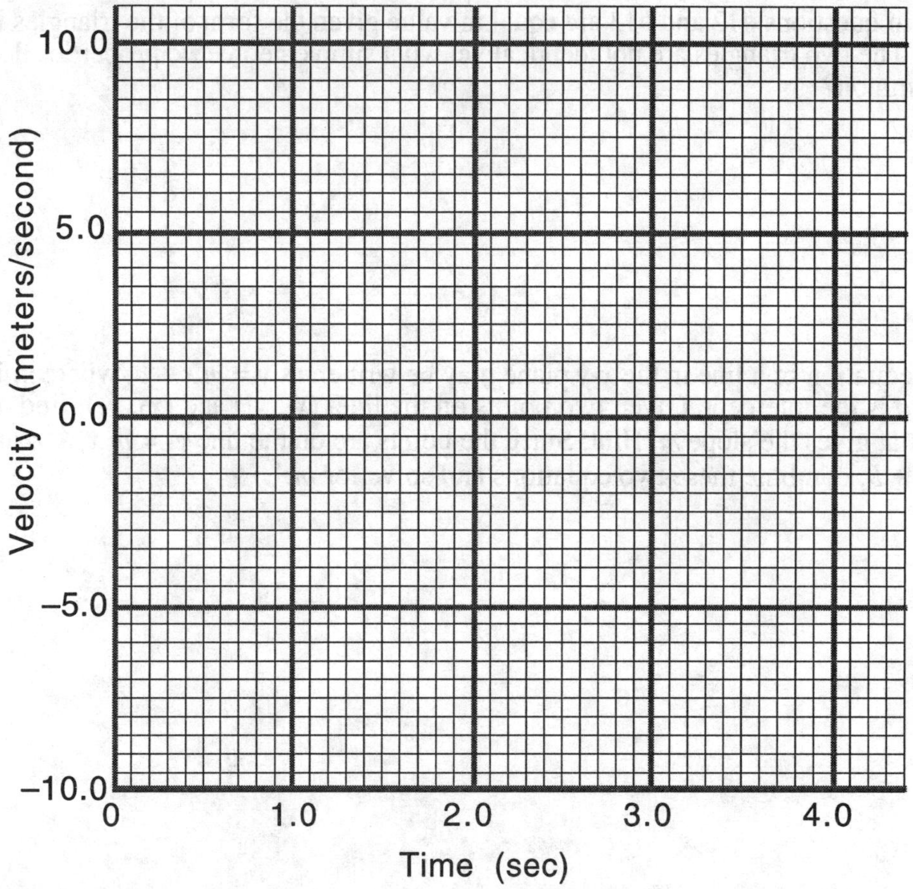

- Graph each data point as a distinct, small dot.
- To increase the visibility of the data, circle each dot, as shown below

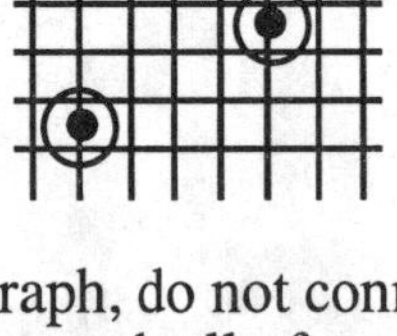

- For this graph, do not connect the data points with lines.
- Be sure to graph all of your data.

Name: ___________________________ Instructor: ___________________

Partner: __________________________ Section Day/Time: ______________

Project 4
Vectors

Objectives

In this project, you observe the collision of two objects and show how the velocity vectors before and after the collision are related. The topics covered in this project include:

- Vector components
- Vector magnitude and direction
- Vector subtraction
- Vector/scalar multiplication

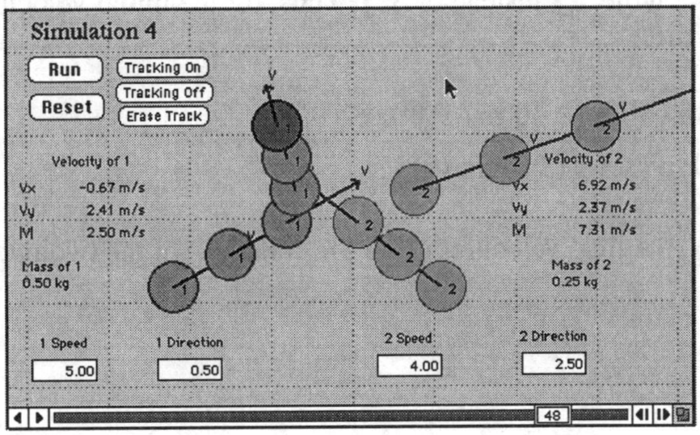

Procedure

Using the *Interactive Physics* program, run the simulation for this project. Two hockey pucks (of unequal mass) slide towards each other on the frictionless surface of an ice rink. The pucks collide and the collision changes their velocities. In this project you will show how the velocities before and after the collision are related; to discover the relationship will require manipulating vectors.

Digital meters measure the x component of velocity, Vx, the y component of velocity, Vy, and the magnitude of velocity, $|V|$, for each puck. Notice that while these values are changed by the collision, they remain constant before and after the collision.

The tracking feature in the simulation allows you to view the "track" of the pucks, somewhat like having a stroboscopic photograph of the motion. Using the push buttons, tracking can be turned on and off and tracks can be erased. To understand the motion of the pucks, you will probably find it helpful to run the simulation with and without tracking.

The initial speed and direction of each puck may be set by typing in the desired value into the appropriate box. Be sure to *reset* the simulation *before* making any changes. The direction angle is entered in radians. The vectors pointing out of the pucks indicate the velocities of the pucks.

Questions and Problems

1. Reset the simulation and set the initial speed of puck 1 to 5.00 m/s and its initial direction to 0.50 radians. Set the initial speed of puck 2 to 4.00 m/s and its initial direction to 2.50 radians. Run the simulation and observe the collision of the pucks. The velocities change with the collision but do they also change as the pucks move apart?

2. There are two ways to manipulate vectors—by components or by magnitude and direction. First we consider working with components. When writing a velocity vector in terms of x and y components, we often use the notation,

$$\vec{\mathbf{V}} = (x\text{ component})\vec{\mathbf{i}} + (y\text{ component})\vec{\mathbf{j}}$$
$$= (Vx)\vec{\mathbf{i}} + (Vy)\vec{\mathbf{j}}$$

Note: Sometimes we write $\hat{\mathbf{i}}, \hat{\mathbf{j}}$ instead of $\vec{\mathbf{i}}, \vec{\mathbf{j}}$. Thus, for the initial velocities of the two pucks we would write,

$$\vec{\mathbf{V}}_{1i} = (4.39\text{ m / s})\vec{\mathbf{i}} + (2.40\text{ m / s})\vec{\mathbf{j}}$$
$$\vec{\mathbf{V}}_{2i} = (-3.20\text{ m / s})\vec{\mathbf{i}} + (2.39\text{ m / s})\vec{\mathbf{j}}$$

What are $\vec{\mathbf{V}}_{1f}, \vec{\mathbf{V}}_{2f}$, the final velocities (after the collision) of the two pucks?

3. Call $\vec{\mathbf{C}}$ the difference between vectors $\vec{\mathbf{B}}$ and $\vec{\mathbf{A}}$, that is, $\vec{\mathbf{C}} = \vec{\mathbf{B}} - \vec{\mathbf{A}}$. Using the components of vectors $\vec{\mathbf{A}}$ and $\vec{\mathbf{B}}$, the components of $\vec{\mathbf{C}}$ are,

$$Cx = Bx - Ax$$
$$Cy = By - Ay$$

with $\vec{\mathbf{C}} = (Cx)\vec{\mathbf{i}} + (Cy)\vec{\mathbf{j}}$. Using components, compute the difference between the final and initial velocities for puck 1, $\vec{\mathbf{D}}_1 = \vec{\mathbf{V}}_{1f} - \vec{\mathbf{V}}_{1i}$.

4. As in question #3, compute the difference between the final and initial velocities for puck 2, $\vec{\mathbf{D}}_2 = \vec{\mathbf{V}}_{2f} - \vec{\mathbf{V}}_{2i}$.

5. From the answers to questions #3 and #4, do you notice any relationship between $\vec{\mathbf{D}}_1$ and $\vec{\mathbf{D}}_2$? Remember that the masses of the pucks are in a ratio of 2:1.

6. The magnitude of a vector may be computed from the components using the Pythagorean theorem (the sum of the square of the sides equals the square of the hypotenuse). For a velocity vector with *x* and *y* components, *Vx* and *Vy*, the magnitude is

$$\left|\vec{\mathbf{V}}\right| = \sqrt{Vx^2 + Vy^2}$$

For example,

$$\left|\vec{\mathbf{V}}_{1i}\right| = \sqrt{(4.39 \text{ m / s})^2 + (2.40 \text{ m / s})^2} = 5.00 \text{ m / s}$$

From the velocities in question #2, determine $\left|\vec{\mathbf{V}}_{2i}\right|$, $\left|\vec{\mathbf{V}}_{1f}\right|$ and $\left|\vec{\mathbf{V}}_{2f}\right|$.

7. From the results of question #3 and #4, determine the magnitudes $\left|\vec{\mathbf{D}}_1\right|$ and $\left|\vec{\mathbf{D}}_2\right|$.

8. The direction of a vector may be computed from its x and y components using trigonometry. The direction angle, θ, relative to the $+x$ axis is

$$\theta = \begin{cases} \text{atan}(Vy / Vx) & \text{if } Vx > 0 \\ \text{atan}(Vy / Vx) + \pi & \text{if } Vx < 0, Vy > 0 \\ \text{atan}(Vy / Vx) - \pi & \text{if } Vx < 0, Vy < 0 \end{cases}$$

where atan is the arctangent function (also known as inverse tangent or $\tan^{-1}$). Notice that this expression gives the angle in radians; be sure your calculator is set to RADIANS. For example, $\theta_{2i} = \text{atan}(\frac{+2.39}{-3.20}) + \pi = 2.50$ radians. From question #3, determine the direction angle for $\vec{\mathbf{D}}_1$.

9. Using questions #4 and #8, determine the direction angle for $\vec{\mathbf{D}}_2$.

10. Convert your answers in questions #8 and #9 from radians to degrees (1 radian = 57.3°). Do you see any relationship between the direction angles of vectors $\vec{\mathbf{D}}_1$ and $\vec{\mathbf{D}}_2$?

11. Vector subtraction is often illustrated graphically. Consider the illustration below,

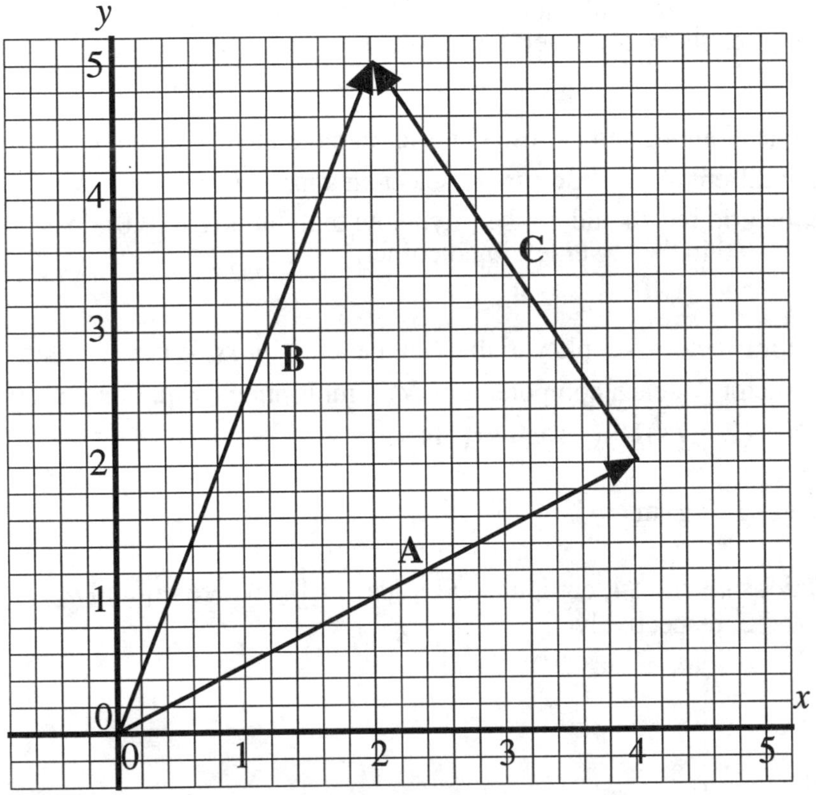

This diagram shows the subtraction $\vec{\mathbf{C}} = \vec{\mathbf{B}} - \vec{\mathbf{A}}$ vectors $\vec{\mathbf{B}} = 2\vec{\mathbf{i}} + 5\vec{\mathbf{j}}$ and $\vec{\mathbf{A}} = 4\vec{\mathbf{i}} + 2\vec{\mathbf{j}}$. Notice that the tails of vectors $\vec{\mathbf{A}}$ and $\vec{\mathbf{B}}$ are together and the vector $\vec{\mathbf{C}}$ goes from the head of $\vec{\mathbf{A}}$ to the head of $\vec{\mathbf{B}}$. Using the components of $\vec{\mathbf{A}}$ and $\vec{\mathbf{B}}$, determine the components of vector $\vec{\mathbf{C}}$; check your answer with the graphical representation of $\vec{\mathbf{C}}$ shown above.

12. From your values in questions #2 and #3, use the grid on page 25 to sketch the vectors $\vec{\mathbf{V}}_{1i}$, $\vec{\mathbf{V}}_{1f}$ and their difference $\vec{\mathbf{D}}_1$. Be sure to select an appropriate scale; your vectors should stay inside the grid but should not be tiny. Note that the origin (0,0) does *not* necessarily have to be located in the lower left hand corner.

13. From your values in questions #2 and #4, use the grid on page 26 to sketch the vectors $\vec{\mathbf{V}}_{2i}$, $\vec{\mathbf{V}}_{2f}$ and their difference $\vec{\mathbf{D}}_2$. Be sure to select an appropriate scale; your vectors should stay inside the grid but should not be tiny. Note that the origin (0,0) does *not* necessarily have to be located in the lower left hand corner.

14. To distinguish them from vectors, simple variables are called *scalars.* When a scalar, m, (like mass) multiplies a vector $\vec{\mathbf{V}}$, each component of $\vec{\mathbf{V}}$ is multiplied by m. For example, if $m = 3$ kg and $\vec{\mathbf{V}} = (2\text{ m/s})\vec{\mathbf{i}} + (-4\text{ m/s})\vec{\mathbf{j}}$ then

$$m\vec{\mathbf{V}} = (6\text{ kg}\cdot\text{m/s})\vec{\mathbf{i}} + (-12\text{ kg}\cdot\text{m/s})\vec{\mathbf{j}}$$

From your results for questions #3 and #4, compute $m_1\vec{\mathbf{D}}_1$ and $m_2\vec{\mathbf{D}}_2$ where m_1 and m_2 are the masses of pucks 1 and 2, respectively.

15. Later in your physics education you will learn that the collision velocities of the pucks may be computed using the principle of *conservation of momentum* and that the vectors $\vec{\mathbf{D}}_1$ and $\vec{\mathbf{D}}_2$ are related as $m_1\vec{\mathbf{D}}_1 = -m_2\vec{\mathbf{D}}_2$. Explain why your answer to question #14 is consistent with this result.

Grid 1: Grid for sketch of the vector difference $\vec{\mathbf{D}}_1 = \vec{\mathbf{V}}_{1f} - \vec{\mathbf{V}}_{1i}$ (question #12).

Grid 2: Grid for sketch of the vector difference $\vec{\mathbf{D}}_2 = \vec{\mathbf{V}}_{2f} - \vec{\mathbf{V}}_{2i}$ (question #13).

Name: ______________________ Instructor: ______________

Partner: ______________________ Section Day/Time: __________

Project 5
Motion on a Curved Path

Objectives

In the previous projects, you studied the motion of objects moving in a straight line. In this project, we consider the motion of an object moving along a curved path, specifically, a pendulum. The topics covered in this project include:

- Trigonometry
- Sine and cosine functions
- Velocity and acceleration in curved motion
- Period and frequency

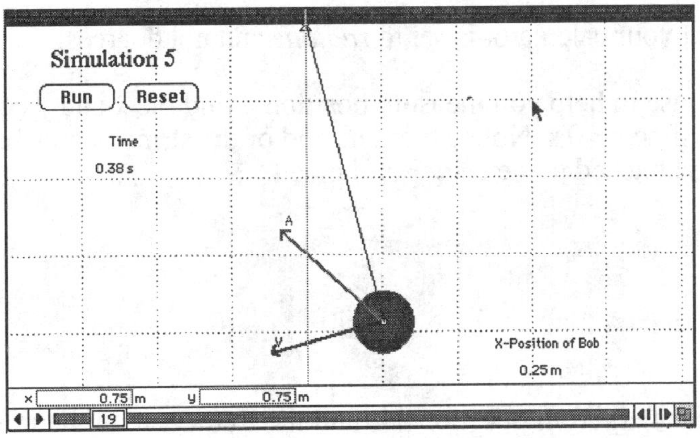

Procedure

Using the *Interactive Physics* program, run the simulation for this project. A simple pendulum, consisting of a ball attached to a string whose other end is fixed to the ceiling, swings back and forth. The ball on the end of the string is called the pendulum "bob." Initially, the pendulum is not moving but the string and bob are at an angle with the vertical.

The velocity and acceleration of the pendulum bob are indicated by vectors (arrows) labeled **V** and **A**, respectively. Initially, the velocity is zero so the velocity vector has zero length. As the bob swings down, its speed increases and the velocity vector lengthens. By definition, the acceleration vector points in the direction of the *change* in the velocity. Notice that at the bottom of the swing, where the bob's motion changes from downward to upward, the acceleration vector points straight up.

You will measure the x coordinate of the pendulum bob and using a bit of trigonometry, compute the angle θ between the string and the vertical (see the figure below). The length of the string (from the center of the bob to the pivot point on the ceiling) is r.

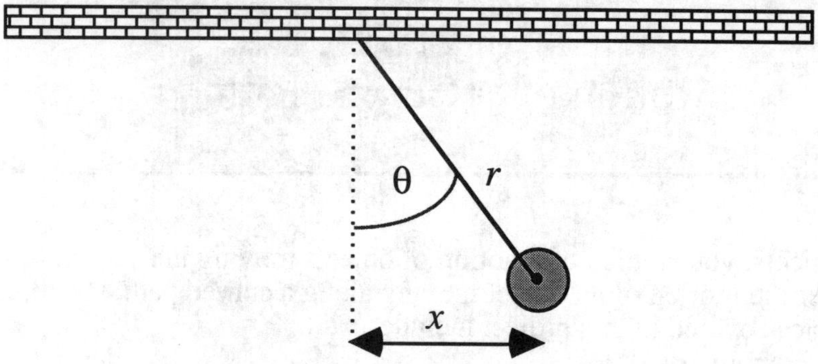

Questions and Problems

NOTE: Be sure that your calculator is set to *radians* and not degrees.

1. Using the mouse to help you measure positions, find the x and y coordinates of each end of the string at time $t = 0$ s. Notice that one end of the string is attached to the ceiling and the other end is attached to the *center* of the ball.

2. Using your answers from question #1, and the Pythagorean theorem, $r^2 = a^2 + b^2$, show that the length of the string is $r = 1.0$ m. Hint: The figure below might help.

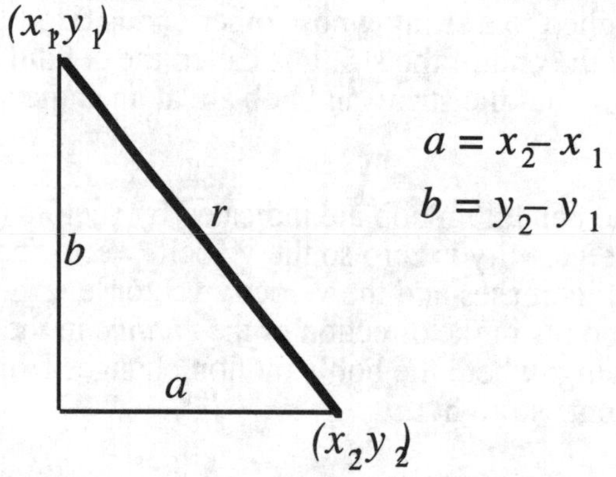

3. The sine of an angle is defined as the ratio of the opposite side to the hypotenuse. For the angle between the string and the vertical, this is $\sin(\theta) = x/r$ (see the figure below).

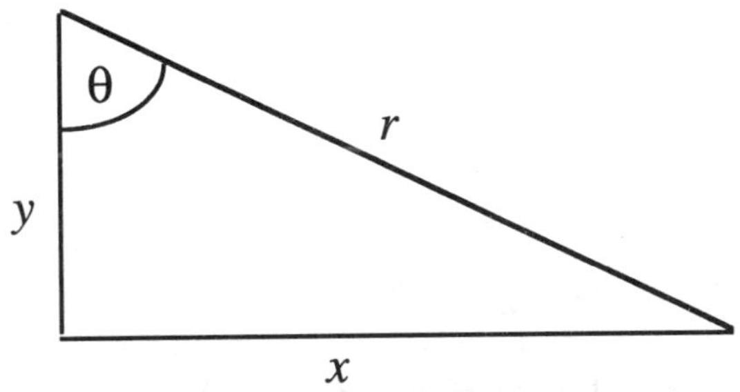

Given x and r we may compute the angle θ using the arcsine function, that is, $\theta = \text{asin}(x/r)$. On some calculators the **asin** key is labeled **sin^{-1}** while on others one obtains arcsine by pressing the **inv** key and then the **sin** key. To test how your calculator works, check the values given below and fill in the missing values. Make sure that your calculator is set to radians.

asin(0.25) = 0.2527   asin(0.75) = ________
asin(0.50) = 0.5236   asin(1.00) = ________

4. Run the simulation, record the x coordinate of the pendulum at various times and fill out column 2 of Table I. Also compute the angle θ using the formula $\theta = \text{asin}(x/r)$ where r (= 1.0 m) is the length of the string (see question #2). Write your values for θ in column 3 of Table I. Again, be sure that your calculator is set to radians.

Table I: Position and angle of pendulum bob at various times.

1 (Data)	2 (Data)	3 (Analysis)	1 (Data)	2 (Data)	3 (Analysis)
t (s)	x (m)	$\theta = \text{asin}(x/r)$	t (s)	x (m)	$\theta = \text{asin}(x/r)$
0.00			2.20		
0.20			2.40		
0.40			2.60		
0.60			2.80		
0.80			3.00		
1.00			3.20		
1.20			3.40		
1.40			3.60		
1.60			3.80		
1.80			4.00		
2.00			4.20		
Graph 1 H-axis		Graph 1 Vertical axis	Graph 1 H-axis		Graph 1 Vertical axis

5. Using the data from Table I, plot the angle θ versus time in Graph 1 (see page 34). Be sure to graph *all* the data points.

6. Periodically, the bob returns to its original position. At approximately what times (give numerical values to the nearest 0.02 seconds) during the simulation does the pendulum return to its original position? Use the simulation's VCR-like controls to help you find these times.

7. The time (in seconds) that it takes the pendulum to make one complete swing (i.e., return to its original position) is called the *period* of the pendulum. From your answer to question #6, what is the period of the pendulum? Try to be accurate to the nearest 0.02 s.

8. At approximately what times is the pendulum bob at zero angle (i.e., hanging straight down so that $x = 0$)? Be accurate and give your answers to the nearest 0.02 seconds. Use the simulation's VCR-like controls to help you find these times.

9. What is the relationship between your answer to question #8 and the period of the pendulum?

10. The linear frequency, f, of the pendulum is related to its period, T, as,

$$f = \frac{1}{T}$$

What is the linear frequency of the pendulum? The units of linear frequency are cycles per second.

11. The angular frequency, ω, of the pendulum is related to the linear frequency by the relation,

$$\omega = 2\pi f$$

Find the angular frequency for our pendulum in radians per second. Note that 2π radians equals one cycle.

12. Your graph of θ versus time should look like a cosine function. It turns out that the angle θ for a pendulum may be estimated using the formula,

$$\theta = \theta_o \cos(\omega t)$$

where θ_o is the initial angle (i.e., θ_o is the angle at $t = 0.0$ s). Using the angular frequency computed in question #11 and θ_o from the first entry in Table I, compute θ using this formula for $t = 0.6$ s. Compare with the value for θ at $t = 0.6$ s in Table I.

13. Using the formula from question #12, find the angle θ for $t = 2.0$ s.

14. Suppose the pendulum swings from right to left. On the figure below, draw in the *velocity* vector, **V**, given by the simulation for each case. Be as accurate as possible in drawing the magnitude and direction of the velocity vector; use the string as a guide.

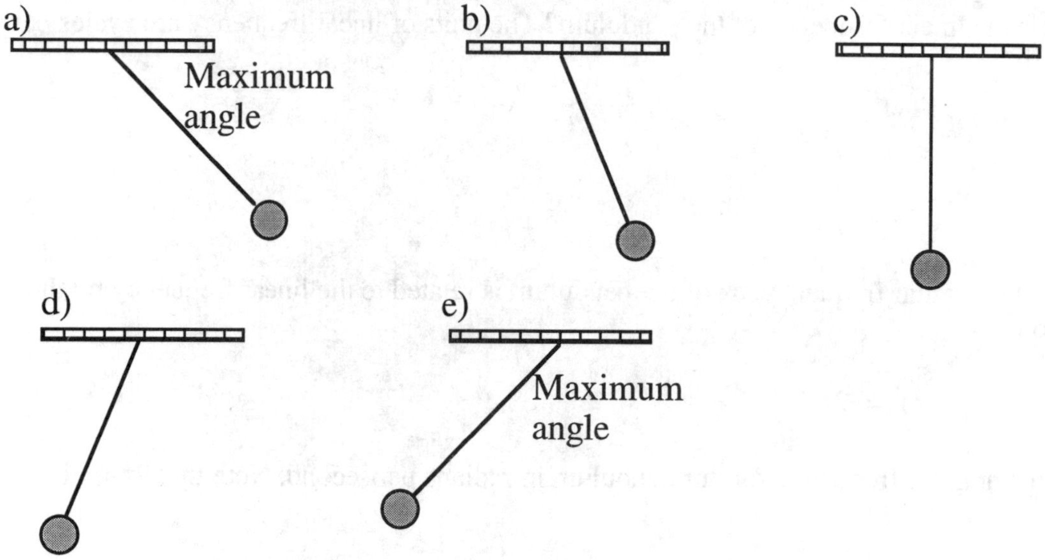

15. At the maximum angle, the velocity is zero. However, the pendulum doesn't stay at that position, rather it swings downward. But how can it move if its velocity is zero? Explain the resolution of this apparent paradox. Hint: Think of another common situation in which the velocity is zero and yet the object does not remain in place.

16. If the pendulum is released at a smaller maximum angle, its velocity as it swings back and forth is certainly smaller yet it turns out that the time it takes the pendulum to complete one swing (the period) stays the same. Explain how this can be if the velocity is smaller.

17. Suppose the pendulum swings from right to left. On the figure below, draw in the *acceleration* vector, **A**, given by the simulation for each case. Be as accurate as possible in drawing the magnitude and direction of the acceleration vector; use the string as a guide.

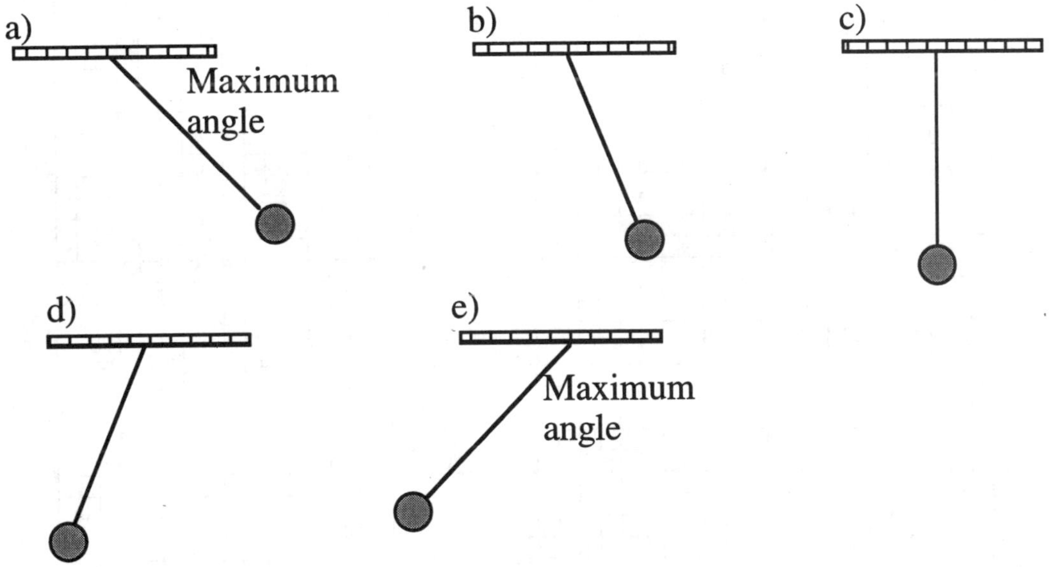

18. At the bottom of the pendulum's swing, the velocity vector points horizontally yet the acceleration vector points straight up. Explain why the acceleration vector points in this direction. Hint: Consider the direction of the velocity vector just before and just after the bottom of the swing and the definition of acceleration.

Graph 1: Angle of pendulum bob versus time

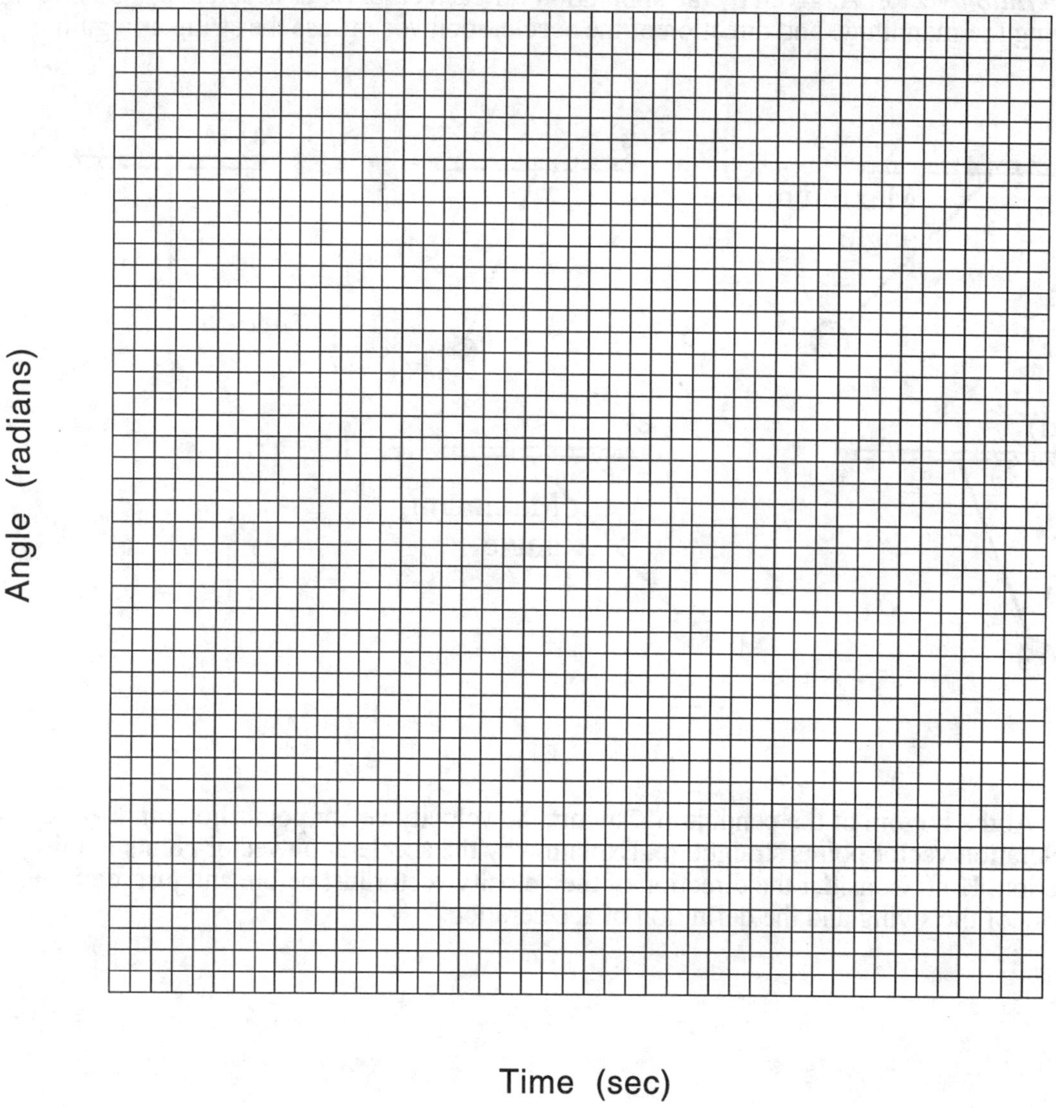

- Before graphing your data, carefully select a suitable scale for the axes; fix your scale so as to fill the page as much as possible.
- Graph each data point as a distinct, small dot.
- To increase the visibility of the data, circle each dot, as shown below

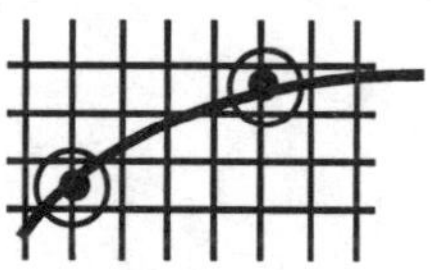

- For this graph, connect the data points with a smooth curve.
- Be sure to graph all of your data.

Name: ______________________________ Instructor: __________________

Partner: ______________________________ Section Day/Time: ____________

Project 6
Velocity and Acceleration I

Objectives

In this project, you observe the motion of an object in one dimension and demonstrate that the acceleration is the derivative of velocity with respect to time. You also study a situation in which two objects are coupled together. The topics covered in this project include:

- Velocity and acceleration
- Derivatives

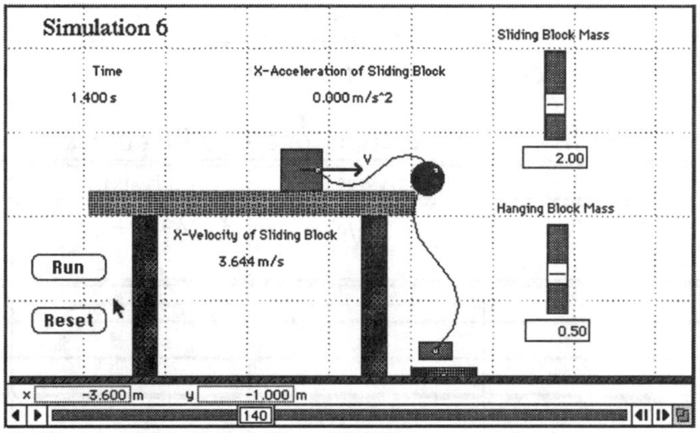

Procedure

Using the *Interactive Physics* program, run the simulation for this project. A large block is sitting on a frictionless table. Attached to this block is a string; the string passes through a pulley. The other end of the string is attached to a smaller block which hangs off the end of the table. The weight of the hanging block can pull the sliding block across the table. There is a rubber pad on the floor to absorb the impact of the falling block.

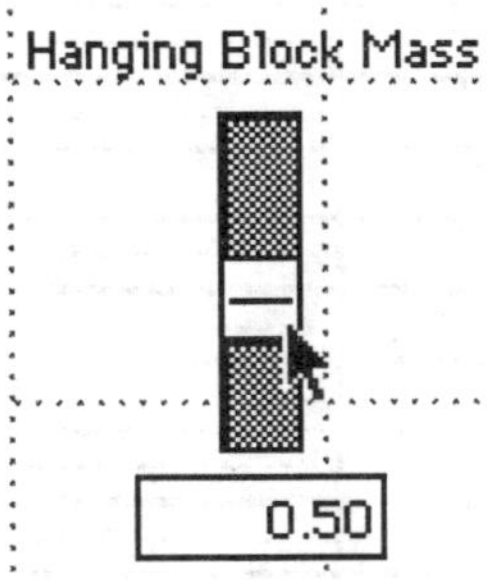

The masses of the two blocks may be adjusted by moving the slider bars up or down (in the picture above, the hanging block has a mass of 0.50 kg). Be sure to reset the simulation before making any changes to the mass of either block.

As the sliding block is pulled across the table by the fall of the hanging block, the speed of the blocks is equal since they must move together (i.e., when the hanging block falls 0.1 m, the sliding block is pulled 0.1 m horizontally). Similarly, the magnitudes of their accelerations are equal while the hanging block is pulling the sliding block across the table. The magnitude of the acceleration turns out to be given by $(m_s + m_h)a = m_h g$ or

$$a = \frac{m_h}{m_s + m_h} g$$

where m_s is the mass of the sliding block, m_h is the mass of the hanging block, a is the magnitude of their acceleration and $g = 9.81$ m/s^2 is the acceleration of gravity. You will be checking this result by varying the masses of the blocks.

Questions and Problems

1. Set the mass of the sliding block to 2.00 kg and the mass of the hanging block to 0.50 kg. Run the simulation and record the velocity and acceleration of the sliding block at various times, filling in Table I. You will be recording your values in columns two and five; the other columns will be computed in later questions.

Table I: Data table of velocity and acceleration for the sliding block.

1 (Data)	2 (Data)	3 (Analysis)	4 (Analysis)	5 (Data)
Time t (seconds)	Velocity v (m/s)	Δv (m/s)	$\Delta v/\Delta t$ (m/s^2)	Acceleration (m/s^2)
0.190		---------------	---------------	---------------
0.200				
0.210		---------------	---------------	---------------
0.390		---------------	---------------	---------------
0.400				
0.410		---------------	---------------	---------------
0.590		---------------	---------------	---------------
0.600				
0.610		---------------	---------------	---------------
0.790		---------------	---------------	---------------
0.800				
0.810		---------------	---------------	---------------
0.990		---------------	---------------	---------------
1.000				
1.010		---------------	---------------	---------------
1.190		---------------	---------------	---------------
1.200				
1.210		---------------	---------------	---------------
1.390		---------------	---------------	---------------
1.400				
1.410		---------------	---------------	---------------
1.590		---------------	---------------	---------------
1.600				
1.610		---------------	---------------	---------------

2. Column 3 of Table I is obtained by taking the difference between velocities from column 2. For example, for the values given in the example below,

Time t (seconds)	Velocity v (m/s)	Δv (m/s)	$\Delta v/\Delta t$ (m/s^2)	Acceleration (m/s^2)
0.190	0.425	---------------	---------------	---------------
0.200	0.445	0.039	1.95	1.962
0.210	0.464	---------------	---------------	---------------

The value in the third column is computed as

$$\Delta v = v(t{=}0.210) - v(t{=}0.190) = (0.464 \text{ m/s}) - (0.425 \text{ m/s}) = 0.039 \text{ m/s}$$

For the next entry you would take the difference between the velocities at times $t = 0.410$ and 0.390, etc. Fill out column 3 for the rest of the Table I.

3. One definition of derivative is

$$\frac{d}{dt} f(t) = \lim_{\Delta t \to 0} \frac{f(t + \tfrac{1}{2}\Delta t) - f(t - \tfrac{1}{2}\Delta t)}{\Delta t}$$

We are going to compute an approximate value for the derivative of velocity, v, by using the formula

$$\frac{dv}{dt} \approx \frac{v(t + \tfrac{1}{2}\Delta t) - v(t - \tfrac{1}{2}\Delta t)}{\Delta t} = \frac{\Delta v}{\Delta t}$$

In the example worked out in question #2, $t = 0.200$ s and $\Delta t = 0.020$ s; the estimated derivative is

$$\frac{\Delta v}{\Delta t} = \frac{0.039 \text{ m/s}}{0.020 \text{ s}} = 1.95 \text{ m/s}^2$$

Finish filling out column 4 in Table I; for *all* entries use $\Delta t = 0.020$ s in your calculation. Your values in column 4 should be almost equal to the values in column 5; if any of them is not, check your work!

4. Explain why the acceleration of the sliding block is zero between 1.3 s and 1.7 s.

5. Reset the simulation and set the mass of the sliding block to 5.00 kg and the hanging block at 0.50 kg. Run the simulation and record the acceleration of the sliding block as it is pulled across the table (before the hanging block hits the floor) in the first entry for column 3 of Table II below. For each entry in the table, reset the simulation and adjust the masses to the various values to finish filling out column 3 in the table.

Table II: Acceleration for various masses of sliding and hanging blocks.

1 (Data)	2 (Data)	3 (Data)	4 (Analysis)
Sliding Block Mass m_s (kg)	Hanging Block Mass m_h (kg)	Acceleration (m/s^2) from computer	Acceleration (m/s^2) from formula
5.00	0.50		
3.00	0.50		
1.00	0.50		
2.00	1.00		
2.00	0.10		
1.00	1.00		

6. Fill out column 4 of Table II using the theoretical formula for the acceleration,

$$a = \frac{m_h}{m_s + m_h} g$$

where m_s is the mass of the sliding block, m_h is the mass of the hanging block (see page 36). Your values in columns 3 and 4 should be similar; if they are not, check your work.

7. From the formula for acceleration given in question #6, what is the acceleration when the mass of the hanging block goes to zero?

8. From the formula for acceleration given in question #6, what is the acceleration when the mass of the sliding block goes to zero?

9. From the formula for acceleration given in question #6, what is the acceleration when the mass of the sliding block is very large, much larger than the hanging block's mass ($m_s >> m_h$)?

10. From the formula for acceleration given in question #6, what is the acceleration when the mass of the hanging block is very large, much larger than the sliding block's mass ($m_h >> m_s$)?

11. How are questions #7 and #9 (and their answers) related?

12. How are questions #8 and #10 (and their answers) related?

13. If the table top is *not* frictionless, then the acceleration of the sliding block turns out to be given by the formula,

$$a = \frac{1}{m_s + m_h}(m_h g + F)$$

where the constant F includes the effect of friction. If the friction is not large so the blocks still move, does F have a positive or a negative value?

14. If the friction is large so the blocks do not move at all, what is the value of F in terms of the masses of the blocks and the gravitational acceleration g?

Name: ______________________________ Instructor: __________________

Partner: ______________________________ Section Day/Time: ____________

Project 7
Velocity and Acceleration II

Objectives

In this project, you observe the motion of a spring oscillator and demonstrate that the velocity is the integral of acceleration with respect to time. The topics covered in this project include:

- Integration
- Velocity and acceleration

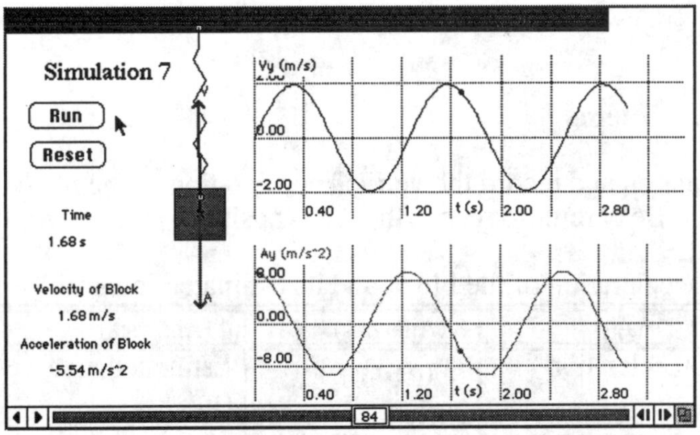

Procedure

Using the *Interactive Physics* program, run the simulation for this project. A spring (represented by the jagged line) has one end connected to the ceiling and the other end attached to a block. Initially, the spring is stretched out and it accelerates the block upward. However, the force exerted by the spring decreases as the spring contracts. At one point, the downward force of gravity equals the upward force exerted by the spring. At that point, the acceleration of the block is zero, however, the block is still moving upward.

Eventually, the velocity goes to zero at the top of the oscillation (when the spring is compressed to its minimum length), the forces of gravity and the compressed spring then push the block back downward. Finally, we return to our original state completing one oscillation; the motion repeats in this fashion.

Two vectors (i.e., arrows) illustrate the velocity and acceleration of the block as it moves up and down. The velocity vector is longest when the block is moving the fastest. The acceleration vector is longest when the velocity is most rapidly changing. The vertical velocity and acceleration of the block are displayed on digital meters and graphed as a functions of time.

In general, the velocity and acceleration are related by the fact that the acceleration is the time derivative of the velocity,

$$a = \frac{dv}{dt}$$

Another relationship between a and v is that the velocity equals the integral of acceleration with respect to time, that is,

$$v(t_f) - v(t_i) = \int_{t_i}^{t_f} a(t)dt$$

where t_i and t_f are the initial and final time. If you are not familiar with this expression, do not worry. This project is designed to help you understand this relationship between velocity and acceleration. In the questions, you are asked to measure the acceleration and using a simple approximation to compute its integral show that the integral matches the velocity. You are also asked to make some observations regarding the oscillating motion.

Questions and Problems

1. Run the simulation and record the vertical acceleration of the block, filling in data column 2 of Table I. Be careful to record the sign (positive or negative) in all cases.

Table I: Vertical acceleration of the block and its estimated integral.

1 (Data)	2 (Data)	3 (Analysis)	4 (Analysis)	5 (Analysis)
time (s)	Acceleration (m/s^2)	$\frac{1}{2}(a+a_{old})$ (m/s^2)	Estimated area (m/s)	Total area (m/s)
0.0		-----------	0.000	0.000
0.1				
0.2				
0.3				
0.4				
0.5				
0.6				
0.7				
0.8				
0.9				
1.0				
1.1				
1.2				
1.3				
1.4				
1.5				
Graph 1 Horiz. axis				Graph 1 Vertical axis

2. In Table I, compute the average of adjacent values of acceleration and put your answer into column 3. Study the example below. BE CAREFUL TO INCLUDE THE MINUS SIGNS WHEN ACCELERATION IS NEGATIVE!

time (s)	Acceleration (m/s^2)	$1/2(a+a_{old})$ (m/s^2)
0.0	9.81	-----------
0.1	8.37	$9.09 = 1/2 ((9.81) + (8.37))$
0.2	4.88	$6.62 = 1/2 ((8.37) + (4.88))$
0.3	0.20	$2.54 = 1/2 ((4.88) + (0.20))$

3. In calculus, you learn that the integral of a function is the "area" under the function. That is, the integral,

$$\int_{t_1}^{t_2} a(t)dt$$

equals the area under the function $a(t)$ between the points t_1 and t_2 (see the figure below).

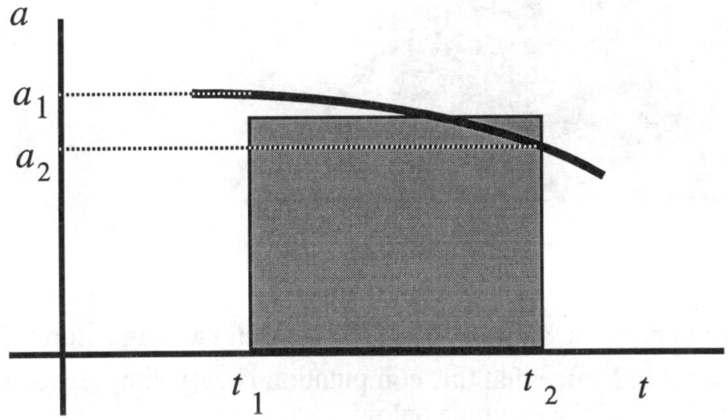

We estimate the area under the curve as the area of the rectangle as,

$$Area = (Base) \times (Height) = (t_2 - t_1)\left(\frac{a_2 + a_1}{2}\right)$$

Suppose $t_1 = 0.1$ s, $t_2 = 0.2$ s, $a_1 = 3.2$ m/s^2 and $a_2 = 5.4$ m/s^2, what is the area? Don't forget the units.

4. Notice that the units in the previous question were *not* square feet or square meters, our customary units of surface area. In calculus, the concept of *Area* is expanded to encompass the general notion of (*Base*)×(*Height*) where *Base* and *Height* may have units different from the units of length. For example, if *Base* has the units of time and *Height* has the units of velocity, what are the units of the *Area*?

5. We are going to estimate the integral of the acceleration as a sum of the areas of many rectangles (see figure below).

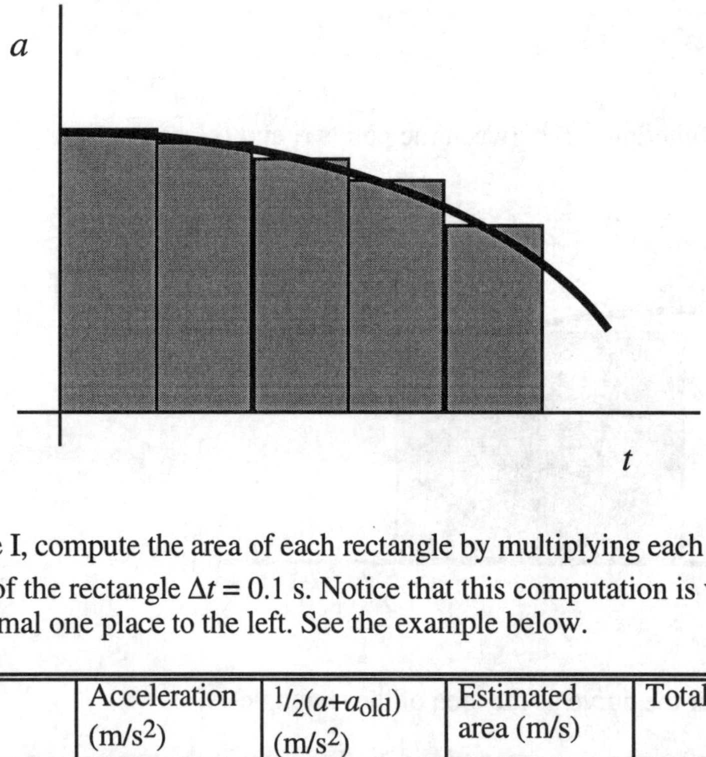

In Table I, compute the area of each rectangle by multiplying each value in column 3 by the "base" of the rectangle $\Delta t = 0.1$ s. Notice that this computation is very simple, just move the decimal one place to the left. See the example below.

time (s)	Acceleration (m/s^2)	$^1/_2(a+a_{old})$ (m/s^2)	Estimated area (m/s)	Total area (m/s)
0.0	9.81	-----------	0.000	0.000
0.1	8.37	9.09	0.909	0.909
0.2	4.88	6.62	0.662	1.571 = (0.909) + (0.662)
0.3	0.20	2.54	0.254	1.825 = (1.571) + (0.254)

6. Finish filling out Table I by computing the running sum of the areas in column 4 and recording the total area in column 5. See the example above. AGAIN, BE CAREFUL WITH THE MINUS SIGNS!

7. Using the data from Table I, graph the estimated integral of the acceleration (column 5) versus time. You may use the graph paper on page 47. Visually compare your graph with the plot of velocity produced by the simulation; the two should be similar, if they differ significantly go back and check your calculations.

8. In the figures below, the block is shown at various positions as it oscillates up and back. In each case, sketch in the *velocity* vector as displayed by the simulation.

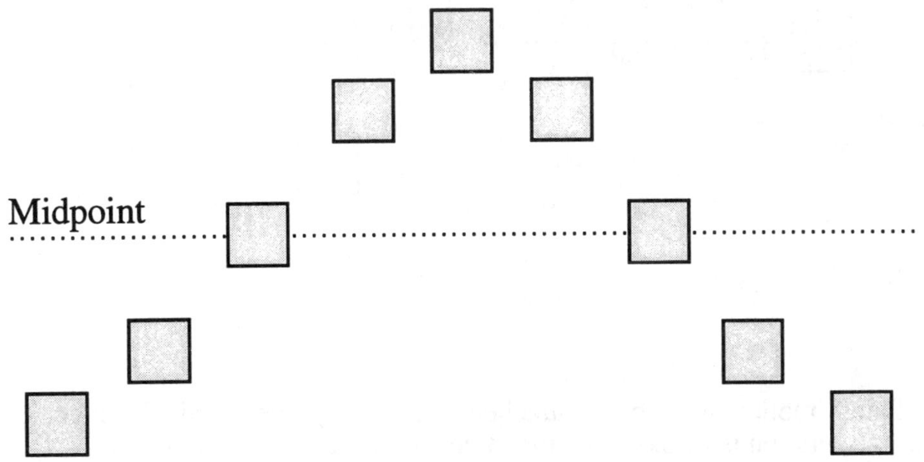

9. The oscillating motion of a block attached to a spring is different from the motion of a bouncing ball. Specifically, as each object moves up and down how does the velocity differ in the two cases?

10. In the figures below, the block is shown at various positions as it oscillates up and back. In each case, sketch in the *acceleration* vector as displayed by the simulation.

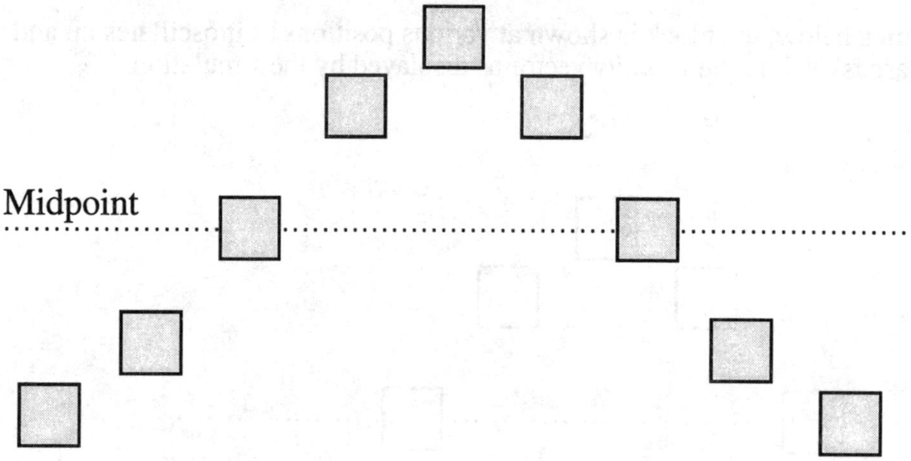

11. In the oscillatory motion of the block attached to the spring, the speed of the block is largest when the acceleration is zero. Is this true in general for any object in motion?

12. In the oscillatory motion of the block attached to the spring, the speed of the block is zero when the acceleration is the greatest. Is this true in general for any object in motion?

13. The oscillating motion of a block attached to a spring is different from the motion of a bouncing ball. Specifically, as each object moves up and down how does the acceleration differ in the two cases?

Graph 1: Estimated integral of the acceleration of the block versus time.

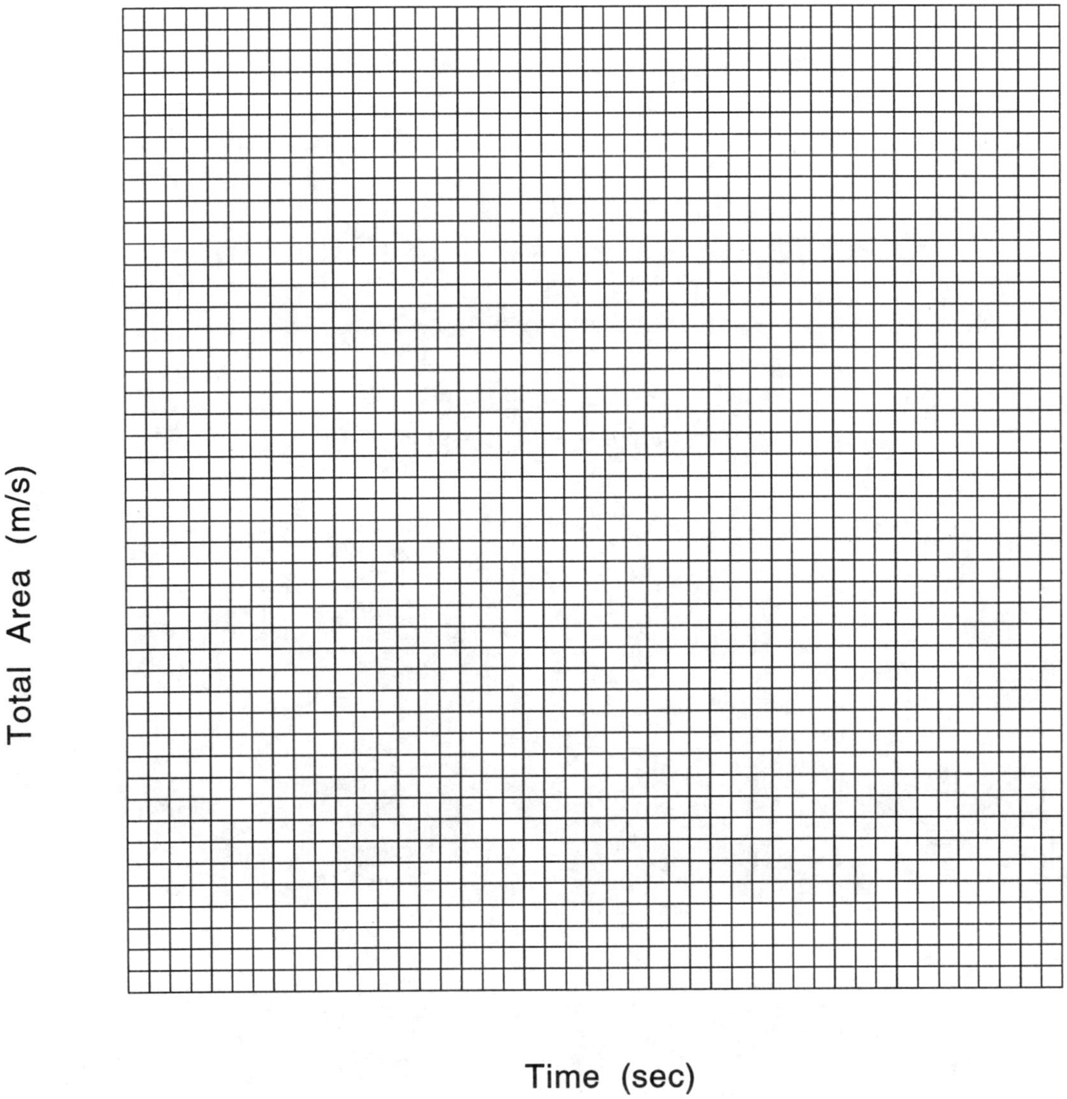

Name: ______________________________ Instructor: ____________________

Partner: ______________________________ Section Day/Time: ____________

Project 8
Projectile Motion

Objectives

In this project, you observe the two dimensional motion of a simple object—a ball tossed at a target. The motion is easy to calculate since we may treat the horizontal and vertical components separately. The topics covered in this project include:

- Vectors
- Projectile motion

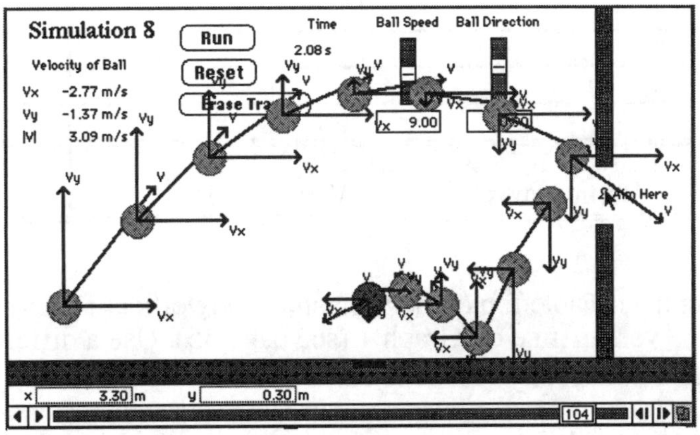

Procedure

Using the *Interactive Physics* program, run the simulation for this project. A rubber ball is thrown towards a target, at a hole in a wall. The ball's shadow can be seen on the ground (the Sun is directly overhead). For some values of speed and direction the ball will pass through the hole; finding those values will be one of your tasks.

The ball's velocity is displayed by the digital meters. The three quantities displayed are the x component of velocity ,V_x, the y component of velocity, V_y, and the magnitude of velocity, $|V|$, which is also called the speed. A vector indicates the ball's velocity; the x and y components of the vector velocity are also displayed.

The ball's initial speed and direction can be adjusted by using the slider bars or by simply typing in a value in the box below the sliders. Be sure that the simulation has been reset *before* making any changes. Try putting in various values for initial speed and direction to see how well you can do at throwing the ball through the hole (then work through the problems to compute these values).

Questions and Problems

1. Run the simulation using a ball speed of $|V| = 9.00$ m/s and direction of $\theta = 0.90$ radians. Record the horizontal, vertical and total velocity at various times, filling in columns 2–4 of Table I. Be careful to record the sign (positive or negative) in all cases.

Table I: Velocity of the ball.

Column 1 (Data)	Column 2 (Data)	Column 3 (Data)	Column 4 (Data)
time (s)	V_x (m/s)	V_y (m/s)	$|V|$(m/s)
0.0			
0.1			
0.2			
0.3			
0.4			
0.5			
0.6			
0.7			
0.8			
0.9			
1.0			
Graph 1 Horizontal axis	Graph 1 Vertical axis	Graph 1 Vertical axis	

2. Using the data from Table I, plot the horizontal and vertical components of the velocity, V_x and V_y, versus time on Graph 1 (see page 55). Use a different symbol to mark the data for the two velocity components.

3. In Graph 1, you should observe that one component of velocity changes while the other is constant. Why is the vertical component different from the horizontal component?

4. In Graph 1, the data for the vertical velocity component fall on a straight line. The slope of this line is the average acceleration; compute its value using the data at $t_1 = 0.0$ s and $t_2 = 1.0$ s with the formula,

$$\text{Average acceleration} = \frac{v(t_2) - v(t_1)}{t_2 - t_1}$$

5. The vertical velocity for an object accelerated by gravity is given by the formula,

$$V_y = V_{0y} - gt$$

where V_{0y} is the object's velocity at time $t = 0$ and $g = 9.81$ m/s^2 is the gravitational acceleration at the Earth's surface. Given that $V_{0y} = 7.07$ m/s, use this equation to compute the vertical velocity of the ball at $t = 1.0$ s. How does your value compare with the value given by the simulation?

6. Repeat question #5 but for $t = 2.0$ s. Why does your value differ from the value given by the simulation?

7. The horizontal and vertical positions of an object in projectile motion are given by the equations,

$$x = x_0 + V_{0x}t$$

$$y = y_0 + V_{0y}t - \tfrac{1}{2}gt^2$$

where x_0, y_0 give the position at time $t = 0$ and V_{0x}, V_{0y} give the horizontal and vertical velocity at time $t = 0$. Given that $x_0 = -3.3$ m, $y_0 = -1.0$ m, $V_{0x} = 5.57$ m/s and $V_{0y} =$ 7.07 m/s, find x and y at time $t = 0.8$ s. How does your value compare with the value given by the simulation?

8. Repeat question #7 but for $t = 2.0$ s. How does your value compare with the value given by the simulation? Why is there a difference?

9. Now let's try to shoot the ball through the hole. The center of the hole is located at $x = 3.3$ m and $y = 0.3$ m (check these values using the coordinates displayed at the bottom of the simulation's window). Suppose we want the ball to be at exactly this position at time $t = 1.0$ s. Given that $x_0 = -3.3$ m, $y_0 = -1.0$ m, use the equations given in question #7 to find V_{0x} and V_{0y}.

10. The relation between the components of the velocity and the magnitude of the velocity (i.e., the speed) is

$$|V| = \sqrt{V_x^2 + V_y^2}$$

From your values of V_{0x} and V_{0y} computed in question #9, find the initial ball speed.

11. The relation between the components of the velocity and the angle of the object's direction is

$$\theta = \text{atan}\left(\frac{V_y}{V_x}\right) = \tan^{-1}\left(\frac{V_y}{V_x}\right)$$

From your values of V_{0x} and V_{0y} computed in question #9, find the initial ball direction. Careful—be sure your calculator is set to *radians*.

12. Check your answers from questions #10 and #11 by running the simulation using these values for the initial speed and direction. Reset the simulation and then type your values directly into the boxes (be sure to hit the return key after entering the value). Run the simulation and see if the ball passes through the center of the hole at time $t = 1.0$ seconds, as predicted. If you miss, check your calculations in questions #9 to #11.

13. Repeat question #9 but with the ball passing through the center of the hole at time $t =$ 0.5 s.

14. Repeat question #10 but with the ball passing through the center of the hole at time $t =$ 0.5 s.

15. Repeat question #11 but with the ball passing through the center of the hole at time $t =$ 0.5 s.

16. Check your answers from questions #14 and #15 by running the simulation using these values for the initial speed and direction. Reset the simulation and then type your values directly into the boxes. Run the simulation and see if the ball passes through the center of the hole at time $t = 0.5$ seconds, as predicted. If you miss, check your calculations in questions #13 to #15.

17. Describe the difference between the trajectories when the ball passes through the hole at a time $t = 1.0$ seconds and 0.5 seconds.

18. Describe what the trajectory will look like if the ball passes through the hole at a time t that is very small.

19. Describe what the trajectory will look like if the ball passes through the hole at a time t that is very large.

20. If the velocity is set such that the time at which the ball reaches the hole is extremely large it turns out that the ball fails to pass through the hole. Explain why.

Graph 1: Horizontal and vertical velocity components of the ball versus time.

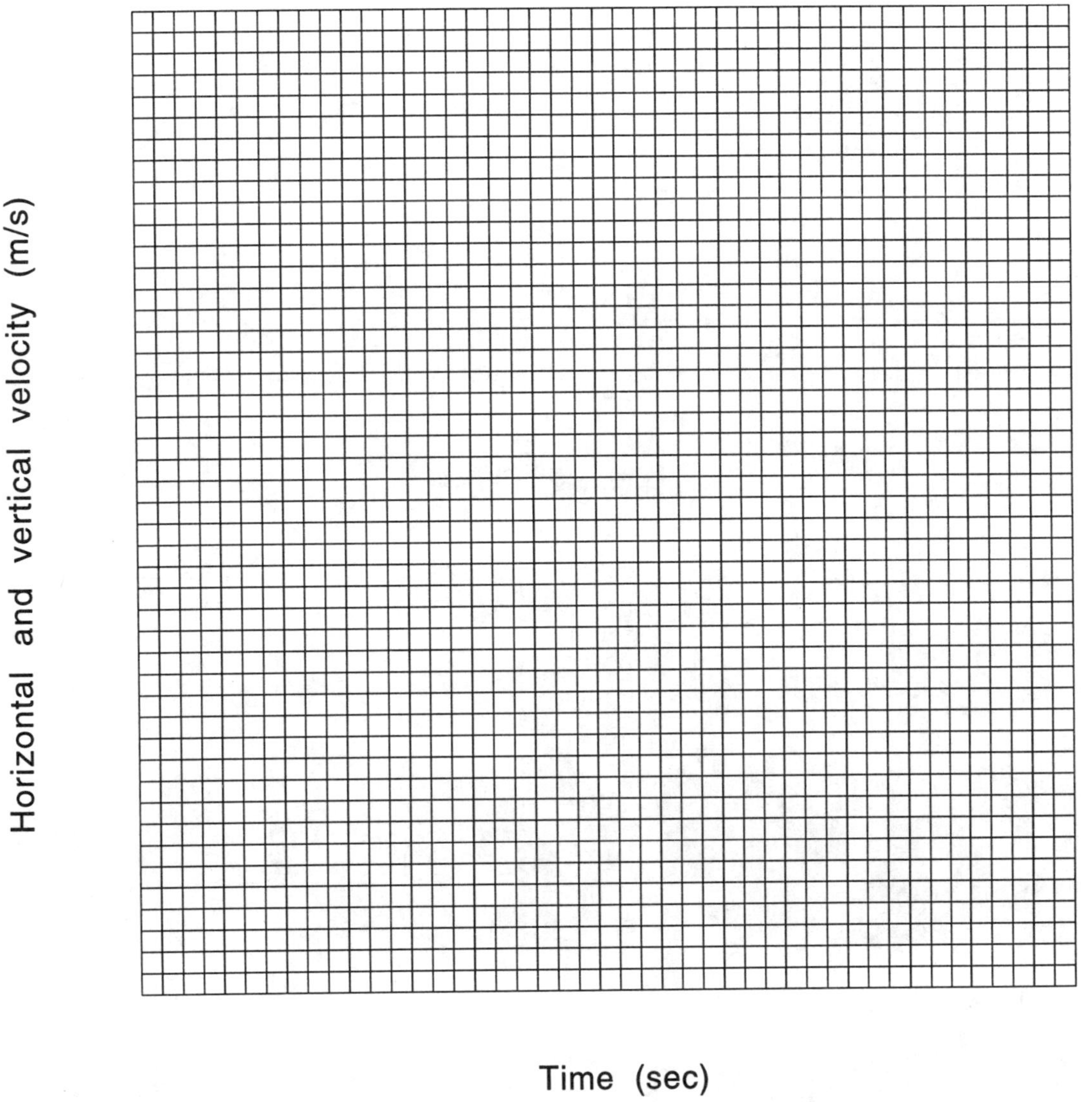

Name: ______________________________ Instructor: ____________________

Partner: ______________________________ Section Day/Time: ____________

Project 9
Forces in One Dimensional Motion

Objectives
In this project, you observe various forces acting on a simple pulley system. The topics covered in this project include:

- Newton's first and second laws
- Tension
- Free-body diagrams

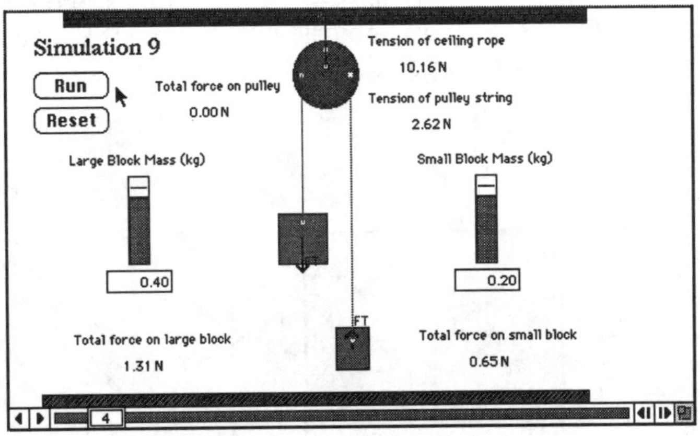

Procedure
In the simulation, two blocks are connected by a string that runs through a pulley. The pulley is attached to the ceiling by a separate rope. Initially, both blocks are above the floor with the larger block (of mass *M*) above the smaller one (of mass *m*); the large block falls, pulling up the small block. When the large block strikes the floor, the string connecting the blocks goes slack. The smaller block continues moving upward but being slowed by the force of gravity, it comes to a halt and then falls downward. After bouncing around for a bit, both blocks eventually come to rest, the large block on the floor and the small block hanging from the string.

There are various forces acting in this system: the gravitational force on the blocks and on the pulley, the tension in the pulley string, the tension in the rope attaching the pulley to the ceiling, the normal force exerted by the floor, etc. The simulation displays the magnitude of some of these forces. The total (or net) force acting on each block is indicated by a vector with the label "FT" next to the vector's arrow head. The masses of the blocks may be changed using the slider controls (be sure to reset the simulation before making any changes using the sliders).

It is important to remember that forces are vectors and, as such, have both magnitudes and directions. The simulation displays the numerical magnitude of the various forces. Since all motion is in the vertical direction, we only need to specify direction as up or down. Instead of using a formal vector notation, we use the simple convention that upward forces are positive and downward forces are negative. For example, the gravitational force on a block is always negative.

Questions and Problems

1. Set the large block's mass to $M = 0.40$ kg and the small block's mass to $m = 0.20$ kg. Run the simulation and observe the motion. Before the large block strikes the floor, what is the total force on the large block? On the small block? If a force is upward, give a positive value while if it is downward give a negative value.

2. The figure below is a free-body diagram indicating the two forces (gravity and the tension in the pulley string) acting on the large block before it hits the floor. The gravitational force is $-Mg$ where M is the mass of the large block and $g = 9.81$ m/s^2 is the gravitational acceleration. On the diagram, indicate which force is which and write in the numerical value (with + or – depending on direction).

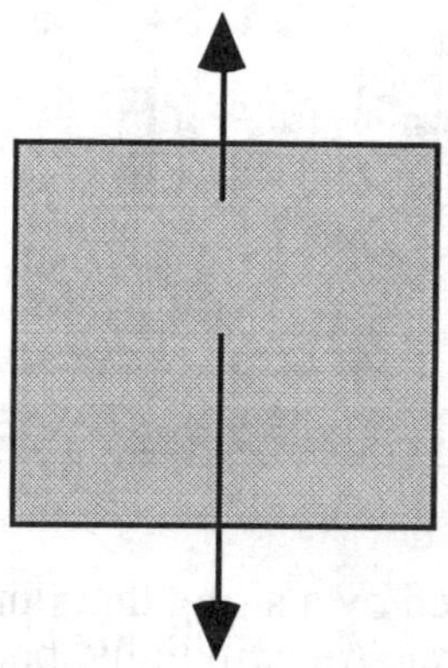

3. The total (or net) force on the large block (before it strikes the ground) is given by

(Net Force) = (Tension) + (Gravity)

or

$$F_{\Sigma} = T - Mg$$

Compute the net force on the large block using this equation. Does your answer agree with the value of the net force displayed by the simulation?

4. By Newton's second law, the force and acceleration are related by $F_{\Sigma} = Ma$. Using the net force given by the digital meter, find the acceleration (in m/s^2) of the large block as it falls. Is the magnitude of this acceleration larger or smaller than g?

5. Draw a free-body diagram, similar to the one in question #2, for the small block (as it is pulled up by the large block).

6. Repeat question #3 but compute the net force on the small block (before the large block strikes the ground).

7. Repeat question #4 but compute the acceleration of the small block (before the large block strikes the ground).

8. From your answers to questions #4 and #7, how are the accelerations of the two blocks related? Why does this relationship exist?

9. Just after the large block strikes the ground, the string goes slack and the small block flies upward freely. Draw the free-body diagram (as in question #2) for the small block.

10. When the string is slack, compare the total force on the small block with its weight, $W = mg$. Why are they related?

11. Run the simulation until the blocks come to rest. When the blocks come to rest, the small block is hanging from the string. The only forces on the block are the tension in the string and the small block's weight. Draw the free-body diagram indicating these forces and including their numerical values.

12. Run the simulation until the blocks come to rest. When the large block is at rest on the ground, the forces acting on it are: the tension in the string, the normal force exerted by the table and the gravitational force on the block (i.e., its weight). On the free-body diagram for the large block shown below indicate which vectors represents each of these forces. Write in the numerical values for the tension and gravitational forces on the diagram.

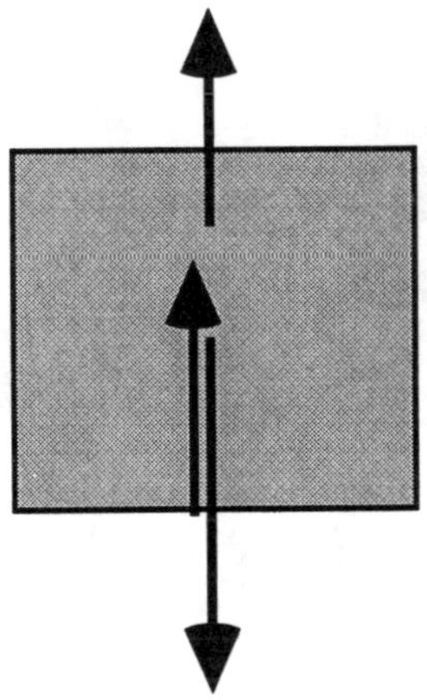

13. Given that the large block comes to rest, the net force on the block must be zero. What is thus the magnitude and direction of the normal force exerted by the table on the block?

14. The figure below is a free-body diagram indicating the forces acting on the pulley. The forces are the tension in the pulley string, the tension in the ceiling rope and the gravitational weight of the pulley itself. On each vector, indicate what force it represents and write in the numerical values for the tensions (rope and string) when the blocks have come to rest. Note that the string hangs off of each side of the pulley.

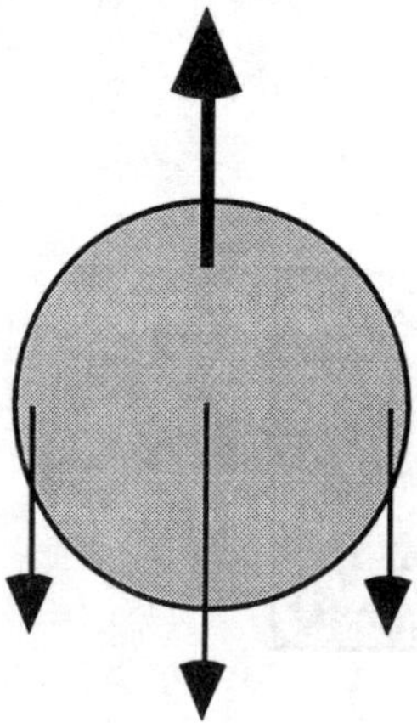

15. The net force on the pulley is

(Net Force) = 2×(String Tension) + (Weight) + (Rope Tension)

The pulley is at rest, so the net force is zero. Using the values for the tension given by the simulation, find the gravitational force on the pulley (i.e., its weight). Be careful, the string tension acts downward while the rope tension acts upward

16. From your answer in question #15, show that the mass of the pulley is 0.5 kg.

17. Suppose we set the mass of each block to 0.2 kg. Work out what the tension in the pulley string will be and check your answer using the simulation.

18. You probably noticed that the tension in the pulley string found in the previous question is the same as the tension in the pulley string when the blocks had different masses and the large block had come to rest on the floor. Explain why the tension is the same in the two cases given that the small block's mass is unchanged.

19. Suppose we set the mass of each block to 0.2 kg. Work out what the tension in the ceiling rope will be and check your answer using the simulation.

20. You probably noticed that the tension in the ceiling rope found in the previous question is the same as the tension in the ceiling rope when the blocks had different masses and the large block had come to rest on the floor. Explain why the tension is the same in the two cases given that the small block's mass is unchanged.

Name: ________________________________ Instructor: ____________________

Partner: _______________________________ Section Day/Time: ____________

Project 10
Forces in Two Dimensional Motion

Objectives
In this project, you observe various forces acting on an object in two-dimensional motion. The topics covered in this project include:

- Newton's laws
- Friction and normal forces
- Free-body diagrams

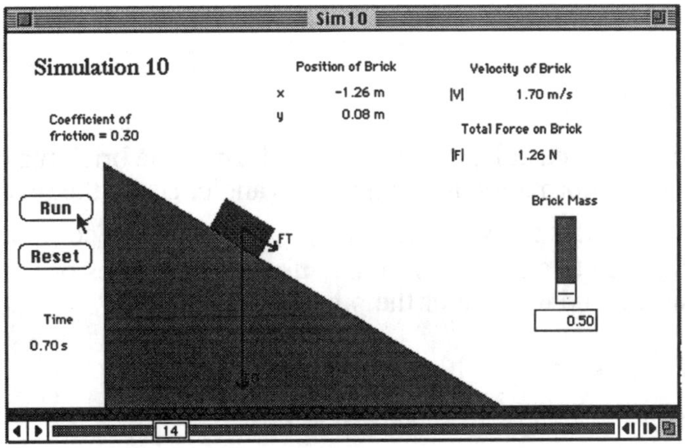

Procedure
Running the simulation, you observe a brick as it is dropped onto and slides down a wooden incline. There is friction between the wooden surface and the brick yet the frictional force is not great enough to keep the brick from moving. Besides the gravitational force on the brick (i.e., its weight), a normal force acts between the wooden incline and the brick. This normal force maintains the brick moving along the surface of the incline. Run the simulation and carefully observe the motion. Concentrate your attention on the motion after the brick is smoothly gliding down the incline, that is, after about 0.5 seconds, when the force on the brick is constant.

Vectors indicate the gravitational force and the total force on the brick; the vectors are marked "FG" and "FT", respectively. As usual, the gravitational force acts in the vertical direction. Notice that the brick's motion is along the incline so the total force must also be parallel to the incline.† The other two forces acting on the brick, the normal force and the frictional force, are not indicated by vectors.

---

† When the brick initially strikes the incline, the total force does not point in the direction parallel to the incline because the wooden surface is somewhat elastic and it exerts a brief impulse on the brick.

The study of this simple system is complicated by the fact that the normal and frictional forces do not act in just the horizontal or vertical direction. The normal force acts perpendicular to the incline while friction acts along the incline. To remove this difficulty we decompose the gravitational force into components parallel and perpendicular to the incline.

The mass of the brick, *m*, may be varied using the slider control. Run the simulation several times with various values of *m*; be sure to press the **Reset** button before setting the value of the mass. You will notice that as the mass increases, the forces increase yet the brick moves with the same velocities and reaches the floor in the same amount of time. You consider this phenomenon in more detail in the questions.

Questions and Problems

1. One of the first things we will need is the angle of the incline. Looking at the incline, we expect the angle is somewhat less than 45°; explain what visual clues indicate this fact.

2. The brick's position is calculated and displayed; since the brick moves along the incline, we may use this information to compute the angle. Using the positions of the brick at times $t = 0.5$ s and $t = 1.5$ s as (x_1,y_1) and (x_2,y_2), respectively, compute the angle θ (see the figure below). Note: In this project, you may use either degrees or radians in measuring angles, just be consistent and use one or the other.

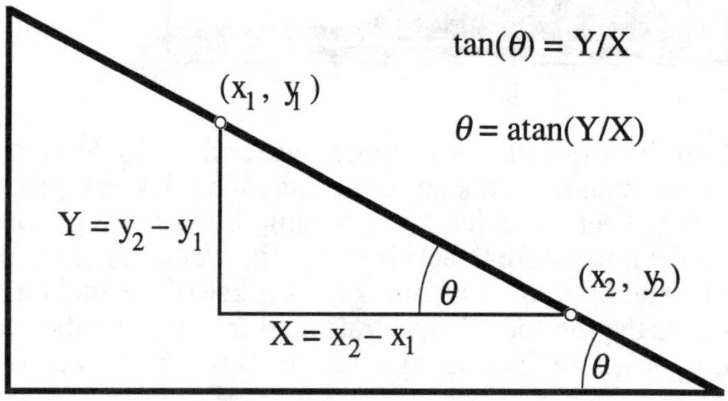

3. Repeat question #2 but using the positions at $t = 0.7$ s and $t = 1.4$ s as (x_1,y_1) and (x_2,y_2), respectively. Show that you get the same answer ($\theta = 31°$) in each case.*

4. Our next step is to draw a free body diagram for the brick. Consider the two free body diagrams shown below. In each case, the three forces acting on the brick are indicated as vectors. We want to decompose the gravitational force into components. Explain why the diagram on the left is incorrect.

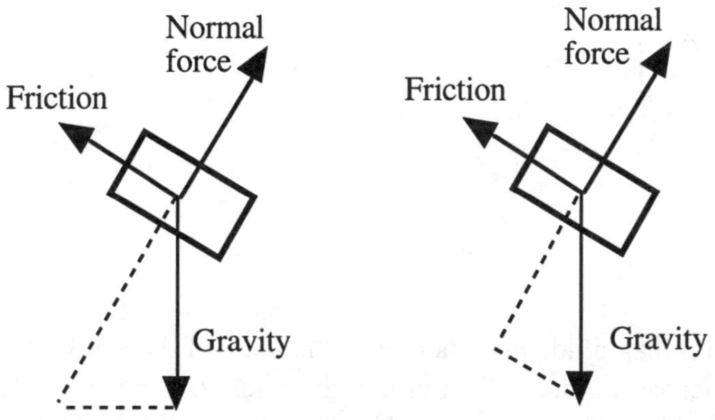

Wrong Right

* Note that we are only interested in the magnitude of this angle, not its quadrant or its sign.

5. To compute the components of the gravitational force we need to determine the angle ϕ in the triangle indicated below. Using basic geometry, find this angle given that you know θ, the angle of the incline. Show your work.

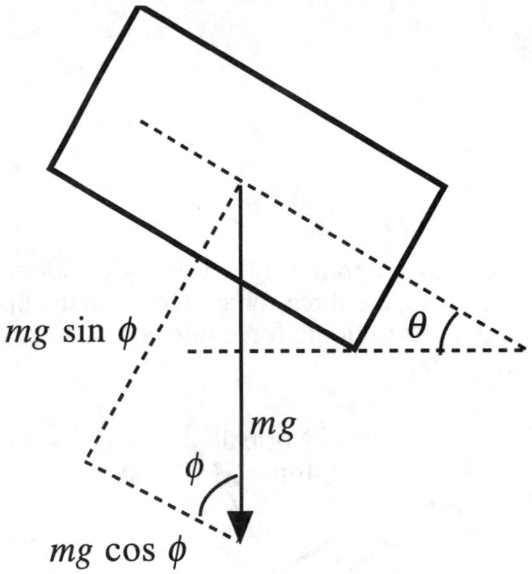

6. The gravitational force has magnitude mg where m is the mass of the brick and g = 9.81 m/s^2 is the gravitational acceleration. Set the mass of the brick to m = 0.5 kg. The component of the gravitational force perpendicular to the incline has a magnitude of $mg \sin \phi$, (see figure above); find the magnitude of this component in Newtons.

7. The magnitude of the normal force, N, must equal the perpendicular component of the gravitational force, $mg \sin \phi$ (see the figure below). Explain why.

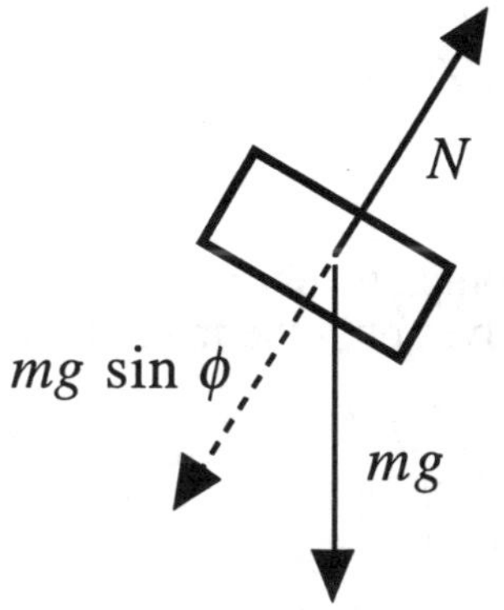

8. The frictional force is related to the normal force as $f = \mu N$ where μ is the coefficient of friction. Given that $\mu = 0.3$ along with your answer in question #6, find the magnitude of the frictional force on the brick in Newtons.

9. The component of the gravitational force parallel to the incline has a magnitude of $mg \cos \phi$; (see the figure in question #5); find the magnitude of this component in Newtons. Again, the brick's mass is $m = 0.5$ kg.

10. There are two forces acting on the brick in the direction parallel to the incline: the gravitational force pulling down the incline and friction acting in the opposite direction, up the incline. The total force is thus,

Total force parallel to the incline = Friction + Gravity

or

$$F_{\Sigma} = \mu N - mg\cos\phi$$

Compute the total force parallel to the incline and compare your value with that given by the simulation when $m = 0.5$ kg (note: the simulation gives only the magnitude of this force).

11. From Newton's second law, $F_{\Sigma} = ma$ where a is the acceleration of the brick down the incline. Using your result from question #10, find this acceleration in m/s^2.

12. The average acceleration, $\overline{a}$, may be computed from the rate of change of velocity with time, that is,

$$\overline{a} = \frac{v_2 - v_1}{t_2 - t_1}$$

where v_1 and v_2 are the velocities at times t_1 and t_2, respectively. Compute the average acceleration using the velocities given by the simulation at times $t_1 = 0.5$ s and $t_2 = 1.0$ s.

13. Repeat question #12 using the times $t_1 = 1.0$ s and $t_2 = 1.5$ s.

14. Reset the simulation and change the brick's mass to $m = 1.5$ kg. Run the simulation and observe the motion of the brick. Does it reach the bottom of the incline sooner, later or at about the same time as before?

15. With $m = 1.5$ kg, as in questions #6 and #7, find the magnitude of the normal force.

16. With $m = 1.5$ kg, as in question #8, find the magnitude of the frictional force.

17. With $m = 1.5$ kg, as in questions #9 and #10, find the total force on the brick parallel to the incline.

18. With $m = 1.5$ kg, as in question #11, find the acceleration of the brick. How does it compare to the previous value computed in question #11?

19. With $m = 1.5$ kg, as in question #12, measure the average acceleration of the brick using the times $t_1 = 0.5$ s and $t_2 = 1.0$ s. How does your answer compare with the previous value computed in question #12?

20. The accelerations you obtained in questions #11, #12, #18 and #19 should all be about the same. Explain, in your own words, why the acceleration of the brick is independent of its mass despite the fact that the gravitational force on the brick increases with mass. If you increase the force on an object, shouldn't that object move faster? Resolve this minor paradox.

Name: ______________________________ Instructor: __________________

Partner: ______________________________ Section Day/Time: ____________

Project 11
Centripetal Force

Objectives

In this project, you observe various forces acting on an object moving in a circle. The topics covered in this project include:

- Centripetal force
- Circular motion
- Newton's laws

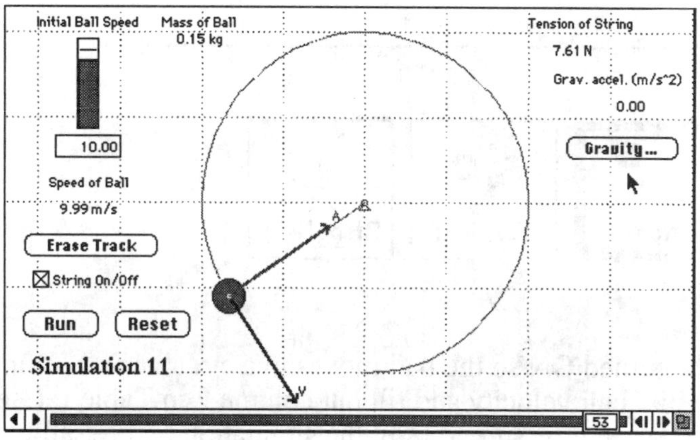

Procedure

In this project, a string connects a ball to a fixed point. The ball has an initial velocity and it moves in a circle (of radius $r = 2.0$ m) around the origin. The tension in the string supplies a force that maintains the radial distance between the ball and the central fixed point to be no greater than the length of the string. The initial ball speed may be set using the slider control. The velocity and acceleration of the ball are indicated by vectors labeled V and A, respectively.

In the simulation, gravity can be turned on and off by pressing the **Gravity** button and selecting either None or Vertical. When gravity is turned off, you should picture the motion of the ball as being in the horizontal plane (for example, sliding on the smooth surface of a lab table). With gravity turned off, because there is no friction, the ball maintains a constant speed as it moves in a circle.

With gravity turned on, imagine the ball moving in the vertical plane, as if you swung it over your head. With gravity turned on, the ball's speed decreases as it moves up to the top of its swing. If the initial speed is low the string may go slack and the ball not trace a complete circle.

The string attached to the ball may be "on" or "cut" by clicking in the String On/Off check box (see below).

☒ String On/Off ☐ String On/Off

String is "on" String is "cut"

This option may be selected while the ball is moving. When the string is cut, the ball is free to move away from the origin.† With gravity turned off, the ball moves in a straight line with constant velocity. With gravity turned on, the ball moves in the parabolic arc of projectile motion.

Questions and Problems

1. Press the **Gravity...** button and check that the gravity option for the simulation is set to None (see below).

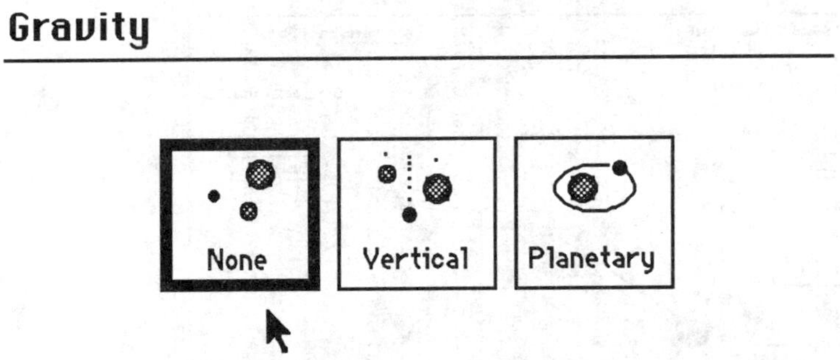

The string should be turned ON so the ball moves in a circular path. Run the simulation for various values of initial ball velocity and fill out column 2 in Table I. Use the slider control to set the ball's initial speed; be sure to reset the simulation before setting the initial speed.

Table I: Tension in the string and centripetal force for various initial ball speeds.

1 (Data)	2 (Data)	3 (Analysis)
Speed (m/s)	Tension (N)	Centripetal force (N)
2.0		
3.0		
4.0		
5.0		
6.0		
7.0		
8.0		
9.0		
10.0		

† Note that even when the string is cut, the simulation draws a faint line between the origin and the ball.

2. The centripetal force is given by

$$F_c = \frac{mv^2}{r}$$

Given that the mass of the ball is $m = 0.15$ kg and the radius of the string is $r = 2.0$ m, compute the centripetal force for each of the velocities in Table I and fill out column 3 in that table.

3. Why is the tension in the string related to the centripetal force? They are equal (except for a bit of computer error) but *why* should they be equal?

4. As observed in the simulation, there is a net force on the ball so from Newton's second law the ball has an acceleration (the acceleration vector is shown in the simulation). Yet the speed of the ball remains constant, shouldn't the speed change due to this acceleration? Describe the resolution of this minor paradox.

5. Reset and run the simulation with the initial velocity set to $v = 10.0$ m/s. Allow the ball to circle around once and then cut the string (press the String On/Off box) just after the ball passes its original position. Sketch the trajectory of the ball.

6. In the figure below, sketch in the velocity and acceleration vectors just before and just after the string is cut.

7. What is the net force acting on the ball after the string is cut?

8. Reset the simulation and using the Gravity button, turn on the vertical gravity (see below) with $a = 9.81\ m/s^2$.

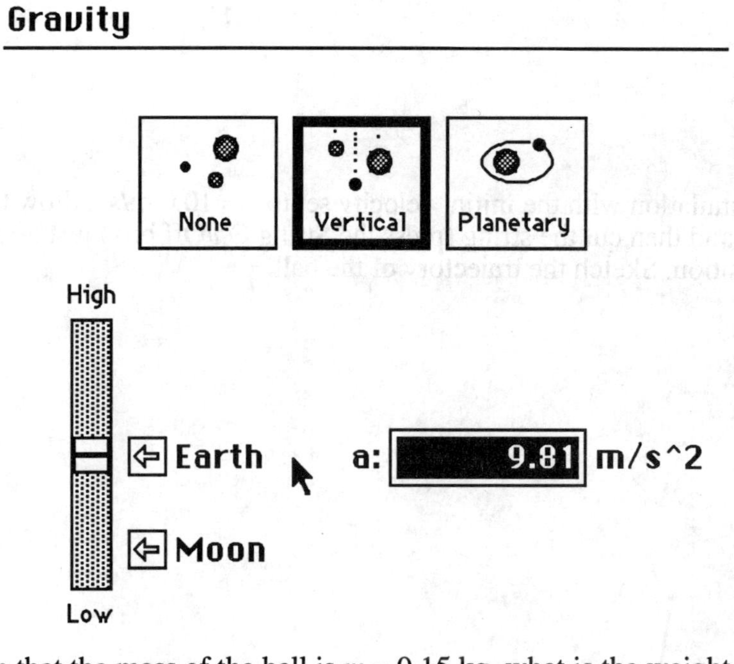

Given that the mass of the ball is $m = 0.15$ kg, what is the weight of the ball (in Newtons)?

9. Run the simulation (be sure the string is ON) for an initial ball speed of $v = 10$ m/s. What is the ball's speed at the top of its swing?

10. Using the formula for centripetal force given in question #2, find F_c when the ball is at the top of its swing.

11. At the top of the swing, the two forces acting on the ball, weight and tension, both act downward (see figure below). How does their sum compare with the centripetal force you computed in question #10 (remember that there is a little bit of computer error in the tension)?

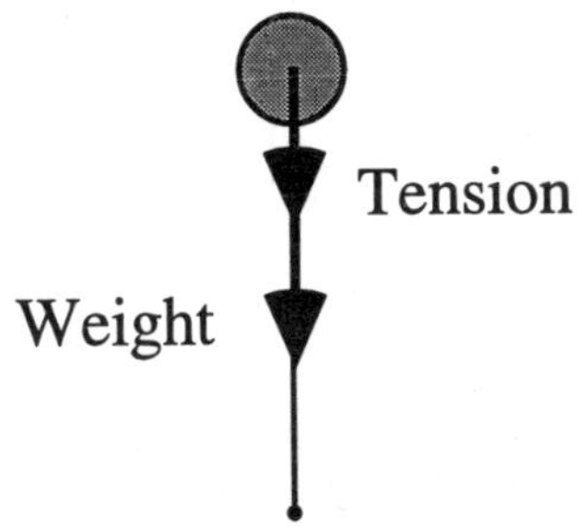

12. Again, run the simulation (be sure the string is ON) for an initial ball speed of $v = 10$ m/s but allow the ball to travel a complete circle. What is the ball's speed at the bottom of its swing?

13. Using the formula for centripetal force given in question #2, find F_c when the ball is at the bottom of its swing.

14. At the bottom of the swing, the two forces acting on the ball, weight and tension, act in opposite directions (the weight acts downward and the tension upward). How does their difference compare with the centripetal force you computed in question #13 (remember that there is a little bit of computer error in the tension)?

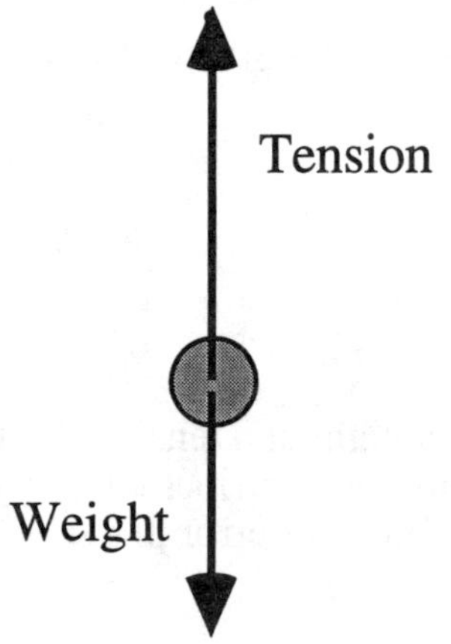

15. The theory of centripetal acceleration states that when an object moves uniformly in a circle, the acceleration is directly inward. With no gravity the acceleration vector always points directly inward however when gravity is turned on this vector only points directly inward at the top and bottom of the circular swing. Explain why the acceleration vector, in general, does not point directly inward when gravity is turned on.

16. With vertical gravity turned on, run the simulation with the initial velocity set to v = 10.0 m/s. Allow the ball to circle around once and then cut the string (press the String On/Off box) just after the ball passes its original position. Sketch the trajectory of the ball.

17. In the figure below, sketch in the velocity and acceleration vectors just before and just after the string is cut.

Before After

18. What is the difference between the trajectories in questions #5 (no gravity) and #16 (vertical gravity)?

Name: ______________________________ Instructor: ____________________

Partner: ______________________________ Section Day/Time: ______________

Project 12
Work and Kinetic Energy

Objectives

In this project, you consider the motion of an object in terms of the work done on that object. Specifically, you observe the direct relationship between the work performed and the change in the object's kinetic energy. The topics covered in this project include:

- Work
- Kinetic energy

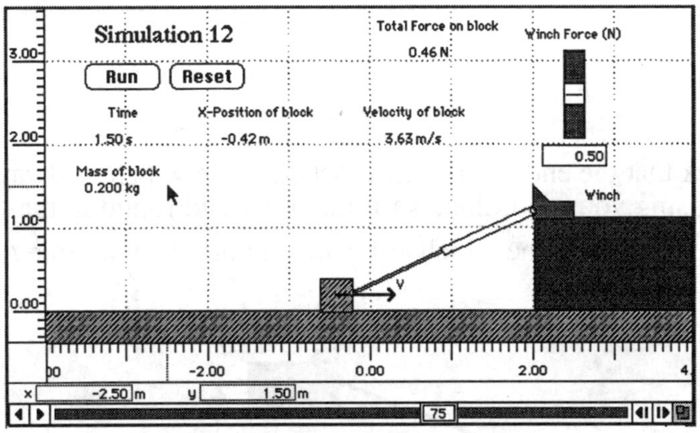

Procedure

In the simulation, a steel block, sitting on the smooth (frictionless) surface of a lab table, is pulled across the table by a winch. The winch exerts a constant force (which may be adjusted using the slider control). The block's speed increases until the block strikes the wooden platform where the winch is attached. The total force on the block (as measured by the digital meter) noticeably changes as the block moves across the table because the angle of between winch arm and the horizontal increases as the block approaches the winch.

The net work done on an object may be computed as

$$\text{Work} = \overline{F}\Delta x$$

where $\overline{F}$ is the average force in the direction of motion.† The displacement, $\Delta x = x_f - x_i$ is the difference between the final and initial positions, x_f and x_i, respectively. The kinetic energy of an object is defined as

$$K = \tfrac{1}{2}mv^2$$

† This average is over distance and not over time. Your textbook probably has a few alternative ways of computing work (e.g., using the integral of the force).

where m is the object's mass and v is the magnitude of its velocity (i.e., its speed). The net work done on an object is equal to the change in the object's kinetic energy, that is,

$$\text{Work} = K_f - K_i$$

where K_f and K_i are the final and initial kinetic energies, respectively. In this project you will directly measure forces, displacements, velocities and energies to confirm this work-energy relation.

Questions and Problems

1. Set the winch force to 0.5 N and run the simulation. At time $t = 0.0$ s, the end of the winch arm attached to the block is located at $x = -3.0$ m, $y = 0.2$ m. NOTE: This is *not* the same as the location of the center of the block since the arm is attached to the side of the block. Find the location of the end of the winch arm attached to the block at time $t = 1.8$ s.

2. You can check that the end of the arm attached to the winch is always located at $x =$ 2.0 m, $y = 1.2$ m. Using your coordinates for the other end found in the previous question, determine the angle θ between the winch arm and the horizontal at time $t = 1.8$ s.

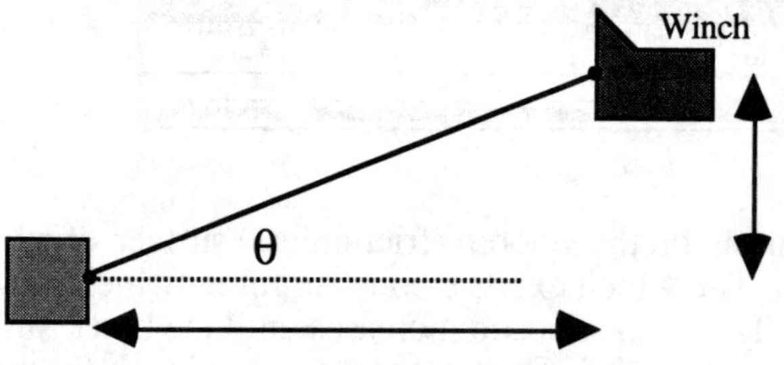

3. Using the angle obtained in the previous question, compute the horizontal component of the winch force, $F_W \cos\theta$, where $F_W = 0.5$ N is the magnitude of the winch force. Compare your value with the total force on the block at time $t = 1.8$ s

4. From the previous questions you should have discovered that the horizontal component of the winch force equals the total force on the block. Explain why the vertical component of the winch force does not contribute to the total force.

5. As the block is pulled across the table the angle of the arm above the horizontal constantly increases so the horizontal component of the force constantly decreases. The velocity of the block increases as it is pulled across the table. Yet if the force is decreasing, shouldn't the block be slowing down? Explain why the block's speed increases.

6. Reset the simulation and run it between time $t_i = 0.0$ s and $t_f = 0.4$ s. Notice that the total force on the block is constant (or nearly so) in this time interval; what is its value?

7. The position of the block is displayed on a digital meter. What is the total displacement of the block, $\Delta x = x_f - x_i$, from time $t_i = 0.0$ s and $t_f = 0.4$ s?

8. For a constant force, the work done by that force equals its magnitude in the direction of motion times the total displacement of the object. Using your results from questions #6 and #7, find the work done on the block from time $t_i = 0.0$ s and $t_f = 0.4$ s. Be sure to not forget units.

9. At time $t_i = 0.0$ s, the object has no kinetic energy since its velocity is zero (i.e., $K_i = 0$ Joules). Using $K = \frac{1}{2}mv^2$, find the final kinetic energy at time $t_f = 0.4$ s. The mass of the block is $m = 0.2$ kg.

10. Compare your results from questions #8 and #9. How are work done and kinetic energy related?

11. When the force is not constant we must compute an average force to determine work. If the force does not vary too much, we can estimate the average force as

$$\overline{F} \approx \frac{F_i + F_f}{2}$$

where F_i and F_f are the initial and final values of the force. Taking the time interval to be $t_i = 1.1$ s and $t_f = 1.5$ s, find $\overline{F}$.

12. Find the displacement of the block, $\Delta x = x_f - x_i$, during the time interval $t_i = 1.1$ s and $t_f = 1.5$ s. Also, compute the work done as $\overline{F}\Delta x$.

13. Compute the initial and final kinetic energies for the initial and final times, $t_i = 1.1$ s and $t_f = 1.5$ s.

14. Compute the change in kinetic energy, $K_f - K_i$, and compare it to the work done on the block (see question #12) in the time interval $t_i = 1.1$ s and $t_f = 1.5$ s.

15. Repeat question #11 for the time interval $t_i = 1.5$ s and $t_f = 1.9$ s.

16. Repeat question #12 for the time interval $t_i = 1.5$ s and $t_f = 1.9$ s.

17. Repeat question #13 for the time interval $t_i = 1.5$ s and $t_f = 1.9$ s.

18. Repeat question #14 for the time interval $t_i = 1.5$ s and $t_f = 1.9$ s.

19. Note that our approximation for the average force was probably not very good since the force varies significantly in the time interval $t_i = 1.5$ s and $t_f = 1.9$ s. Suppose we estimate the average force in this interval by averaging the values of the force at time 1.5 s, 1.6 s, 1.7 s, 1.8 s and 1.9 s. Compute the average of the force at these five values (i.e., sum the five values and divide by five).*

20. Using the average force computed in question #19, compute the work done in the time interval $t_i = 1.5$ s and $t_f = 1.9$ s using $\overline{F}\Delta x$ (see question #16). Compare with the change in kinetic energy computed in question #18.

21. For a winch force of $F_W = 0.5$ N, find the kinetic energy of the block just before it strikes the platform where the winch is attached.

22. Reset the simulation and change the winch force to $F_W = 1.0$ N. Find the kinetic energy of the block just before it strikes the platform where the winch is attached.

23. Describe the relationship between your results from questions #21 and #22.

---

* Note that this is still a crude estimate. We should really be averaging values that are equidistant in space rather than time and treating the average using an integral. Still it is a better estimate than before.

Name: ______________________________ Instructor: ____________________

Partner: ______________________________ Section Day/Time: ______________

Project 13
Conservation of Energy

Objectives

In this project, you observe the motion of two coupled objects and compute the kinetic and potential energy for each object. The topics covered in this project include:

- Gravitational potential energy
- Conservation of energy

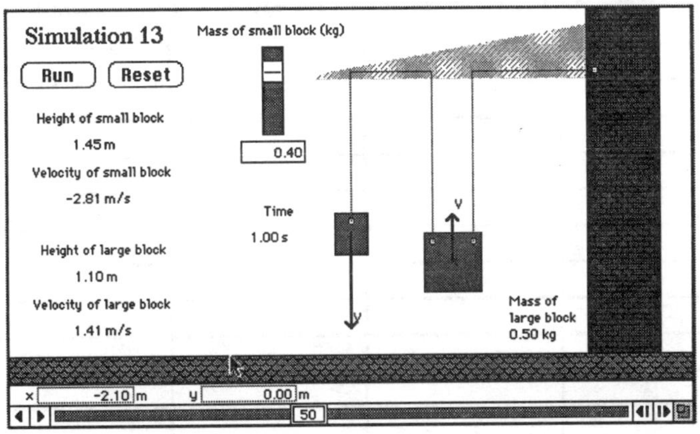

Procedure

Using the *Interactive Physics* program, run the simulation for this project. A frictionless pulley system lifts a large block using a small block as a counterweight. One end of the system's string is attached to the small block. The string runs freely through a set of pulleys attached to a beam and to the large block. The opposite end of the string is tied to a rigid wall.

The vertical position and velocity of each block is displayed by the digital meters. The mass of the small block may be adjusted using the slider control. If the mass of the small block is less than half the mass of the large block then the large block is not lifted. However, if the small block's mass is greater than 0.25 kg (the mass of the large block is 0.50 kg) then the large block is lifted as the small block falls. The greater the mass of the small block, the faster the large block is lifted. The small block bounces when it strikes the ground but eventually the system comes to rest.

The gravitational potential energy of an object located at a height y is given by,

$$U = m g y$$

where m is the mass of the object and $g = 9.81$ m/s^2 is the gravitational acceleration. The height of an object is measured from the center of mass of the object to a reference level (in this case, the floor).

The kinetic energy of an object is

$$K = {}^1/_2\, m\, v^2$$

where v is the magnitude of the object's velocity (i.e., its speed).

Questions and Problems

1. Reset and run the simulation using a small block mass of $m_S = 0.4$ kg. Record the vertical height of the small block, y_S, and large block, y_L at various times, filling in columns 2 and 4 of Table I.

Table I: Potential energy for the blocks.

1 (Data)	2 (Data)	3 (Analysis)	4 (Data)	5 (Analysis)	6 (Analysis)
time (s)	y_S (m)	U_S (Joules)	y_L (m)	U_L (Joules)	U_T (Joules)
0.0					
0.2					
0.4					
0.6					
0.8					
1.0					
1.2					
1.4					
1.6					
1.8					
2.0					
2.2					

Table II: Kinetic energy for the blocks.

1 (Data)	2 (Data)	3 (Analysis)	4 (Data)	5 (Analysis)	6 (Analysis)
time (s)	v_S (m/s)	K_S (Joules)	v_L (m/s)	K_L (Joules)	K_T (Joules)
0.0					
0.2					
0.4					
0.6					
0.8					
1.0					
1.2					
1.4					
1.6					
1.8					
2.0					
2.2					

2. Using the data from Table I, compute the gravitational potential energy for each block using the formula,

$$U_S = m_S\, g\, h_S \quad \text{and} \quad U_L = m_L\, g\, h_L$$

and fill in columns 3 and 5 of that table. The mass of the large block is $m_L = 0.5$ kg.

3. The total potential energy for the two blocks, U_T, is the sum of the potential energies of the individual blocks, that is, $U_T = U_S + U_L$. Fill column 6 in Table I by computing U_T.

4. Keeping the small block mass to be $m_S = 0.4$ kg, record the vertical velocity of the small block, v_S, and large block, v_L at various times, filling in columns 2 and 4 of Table II (see page 88).

5. Using the data from Table II, compute the kinetic energy for each block using the formula,

$$K_S = {}^1\!/_2\, m_S v_S^2 \quad \text{and} \quad K_L = {}^1\!/_2\, m_L v_L^2$$

and fill in columns 3 and 5 of that table. The mass of the large block is $m_L = 0.5$ kg.

6. The total kinetic energy, K_T, is the sum of the kinetic energy of the two blocks, that is, $K_T = K_S + K_L$. Fill in column 6 in Table II by computing K_T.

Table III: Energies for the blocks.

1 (Data)	2 (Analysis)	3 (Analysis)	4 (Analysis)
time (s)	U_T (Joules)	K_T (Joules)	E (Joules)
0.0			
0.2			
0.4			
0.6			
0.8			
1.0			
1.2			
1.4			
1.6			
1.8			
2.0			
2.2			
Graph 1 Horizontal axis	Graph 1 Vertical axis	Graph 1 Vertical axis	Graph 1 Vertical axis

7. Copy column 6 from Table I into column 2 of Table III and column 6 from Table II into column 3 of Table III. Compute the total energy for the two blocks as, $E = U_T + K_T$, and write your values in column 4 of Table III.

8. Using the data in Table III, plot the total potential energy, total kinetic energy and total energy versus time on Graph 1 (see page 92). You should find that the total energy remains approximately constant until the small block strikes the ground.

9. Reset the simulation and change the small block's mass to $m_S = 0.26$ kg. Compute U_T, K_T and E at time $t = 0.0$ s, that is, the initial values of total potential, total kinetic and total energy before the blocks start to move.

10. Run the simulation with $m_S = 0.26$ kg and find U_T, K_T and E just before the small block strikes the ground. Your total energy should be similar to the value obtained in the previous question.

11. Continue running the simulation with $m_S = 0.26$ kg and find U_T, K_T and E after the blocks have come to rest. Is the final value of the total energy E about the same as the initial value computed in question #9?

12. With a small block mass of 0.4 kg, after the blocks come to rest the final value of the total energy is significantly less than the total energy at time $t = 0.0$ s (see Table III and Graph 1). Explain why there is a smaller difference between initial and final values of total energy when the small block's mass is 0.26 kg (see questions #9 and #11).

13. After the blocks come to rest, what is the total distance that the small block has fallen and what is the total distance the large block has been raised? How are these values related?

14. What is the tension in the string when the blocks come to rest? This tension does *not* depend on the mass of the small block. Hint: Consider the forces acting on the large block.

Graph 1: Total potential energy, total kinetic energy and total energy versus time.

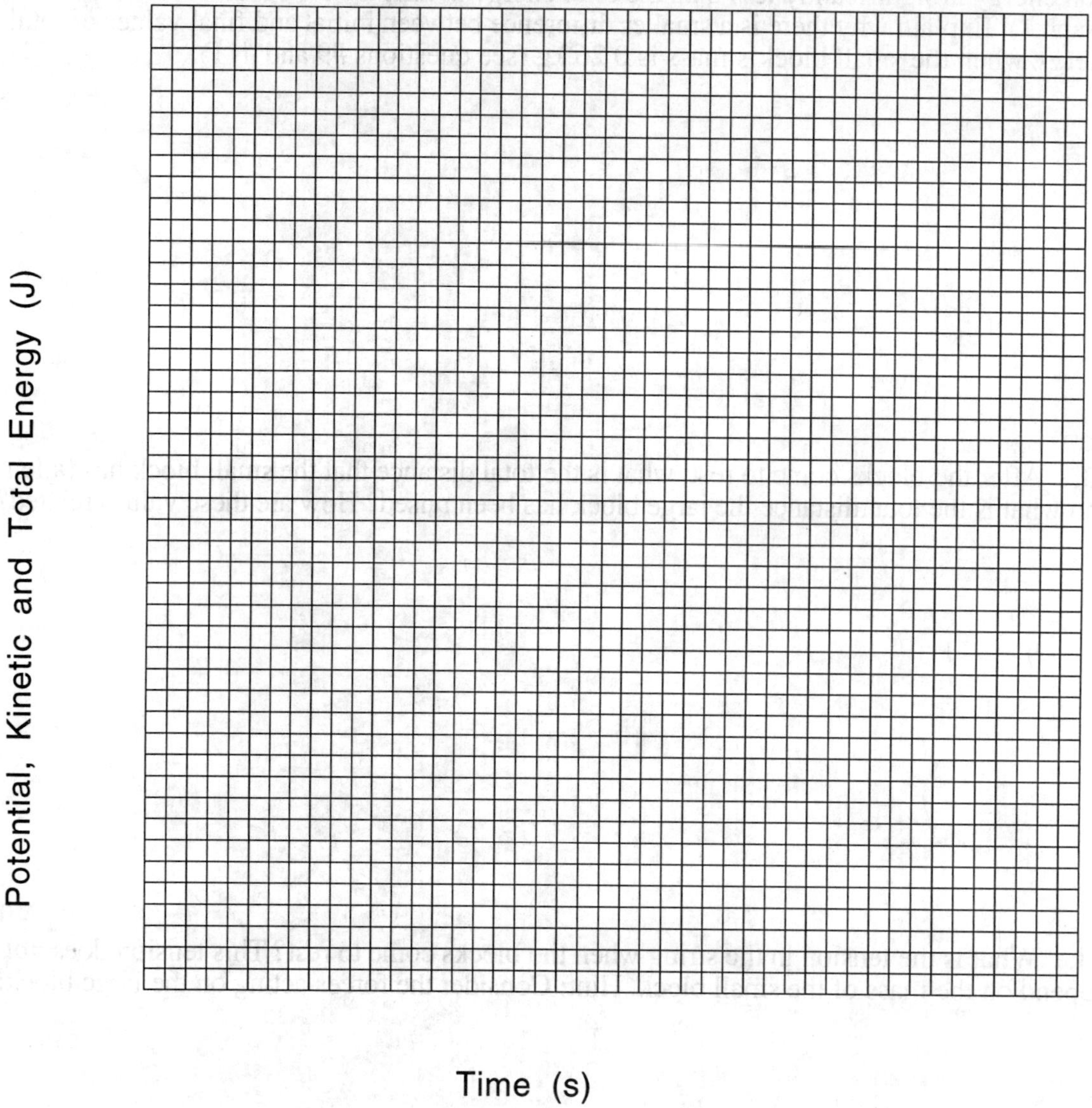

Name: ______________________________ Instructor: ____________________

Partner: ______________________________ Section Day/Time: ____________

Project 14
Momentum and Center of Mass

Objectives

In this project, you observe the motion of two objects in a collision and compute their individual and total momenta. You also locate the center of mass for the two objects and relate its velocity to the total momentum. The topics covered in this project include:

- Momentum
- Collisions
- Center of mass

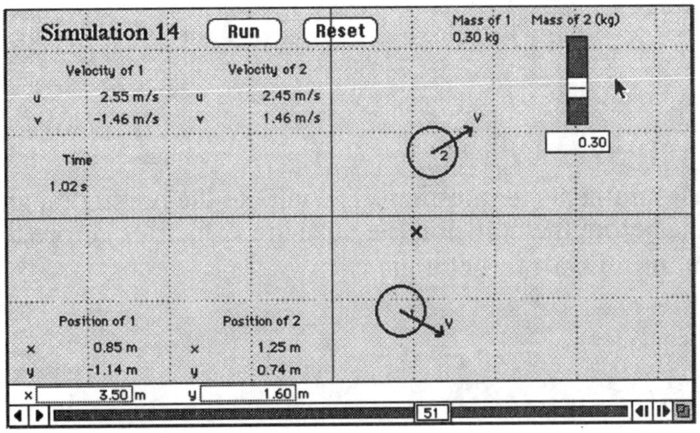

Procedure

Using the *Interactive Physics* program, run the simulation for this project. A clear glass hockey puck slides on a frictionless surface of ice. The puck strikes a second, similar puck and the two move off at an angle after the collision. The vertical position and velocity of each puck is displayed by the digital meters. The mass of the second puck may be adjusted using the slider control.

The location of the center of mass for the two pucks is marked by an **X** on the surface of the ice; this location changes as the pucks move. The pucks are transparent so you can see the **X** mark even if it happens to be under one of the pucks. When the mass of the second puck is small, the center of mass for the two pucks is located closer to the first puck.

The horizontal and vertical momenta for the two pucks is given by the formulas,

$$p_x = m_1 u_1 + m_2 u_2 \qquad \text{and} \qquad p_y = m_1 v_1 + m_2 v_2$$

where m is the mass of a puck; u and v are its horizontal and vertical velocities, respectively.

The horizontal and vertical locations of the center of mass are given by,

$$x_{cm} = \frac{1}{M}(m_1 x_1 + m_2 x_2) \text{ and } \qquad y_{cm} = \frac{1}{M}(m_1 y_1 + m_2 y_2)$$

where M is the total mass, $M = m_1 + m_2$.

Questions and Problems

1. Reset the simulation and set the mass of puck 2 to be $m_2 = 0.3$ kg; notice that the mass of puck 1 is always $m_1 = 0.3$ kg. Run the simulation and measure the horizontal and vertical velocities of each puck before and after the collision.

2. From your measurements in question #1, compute the horizontal and vertical momenta, p_x and p_y, before the collision (i.e., use the velocities before the collision in the formula for momentum). Don't forget units.

3. From your measurements in question #1, compute the horizontal and vertical momenta, p_x and p_y, after the collision. By conservation of momenta, your values of momentum should be equal to those computed in question #2; if not, check your work.

4. The total kinetic energy for the two pucks is given by the formula

$$K = \tfrac{1}{2}m_1\left(u_1^2 + v_1^2\right) + \tfrac{1}{2}m_2\left(u_2^2 + v_2^2\right)$$

Using the velocities from question #1, determine the kinetic energy <u>before</u> the collision.

5. Compute the total kinetic energy <u>after</u> the collision. Note: the collision is *not* elastic so kinetic energy is not conserved.

6. Reset the simulation and set the mass of puck 2 to be $m_2 = 0.6$ kg. Run the simulation and repeat question #1.

7. For $m_2 = 0.6$ kg, use the velocities measured in the previous question to compute the horizontal and vertical momenta, p_x and p_y, <u>before</u> the collision (i.e., repeat question #2).

8. Explain why the values of the horizontal and vertical momenta before the collision turn out to be the same in questions #2 and #7.

9. For $m_2 = 0.6$ kg, use the velocities measured in the question #6 to compute the horizontal and vertical momenta, p_x and p_y, after the collision. By conservation of momentum, your values of momenta should be equal to those computed in question #7; if not, check your work.

10. Reset the simulation and set the mass of puck 2 back to the value of $m_2 = 0.3$ kg. Run the simulation and observe the center of mass of the two pucks (marked with an **X**). For the initial positions of the pucks (i.e., x and y for each puck at time $t = 0.0$ s), compute the expected location of the center of mass using the formulas,

$$x_{cm} = \frac{1}{M}(m_1 x_1 + m_2 x_2) \text{ and } \qquad y_{cm} = \frac{1}{M}(m_1 y_1 + m_2 y_2)$$

where M is the total mass, $M = m_1 + m_2$.

11. Repeat question #10 but find the location of the center of mass at time $t = 0.5$ s.

12. Repeat question #10 but find the location of the center of mass at time $t = 1.0$ s.

13. From your values in questions #10, #11 and #12, compute net horizontal distance traveled by the center of mass from time $t = 0.0$ s to time $t = 0.5$ s. Also compute net horizontal distance traveled by the center of mass from time $t = 0.5$ s to time $t = 1.0$ s.

14. The center of mass moves with a constant velocity. From your values in question #13, what is the horizontal velocity of the center of mass?

15. The horizontal velocity of the center of mass, u_{cm}, may be determined from the total momentum, specifically, $u_{cm} = p_x/M$. Use this formula to compute u_{cm} and compare your answer with the result from the previous question. Recall that you already computed p_x in question #2.

16. The vertical velocity of the center of mass is given by the vertical momentum, specifically, $v_{cm} = p_y/M$. Compute v_{cm} using the value of p_y from question #2.

17. Reset the simulation and set the mass of puck 2 to the value of $m_2 = 0.6$ kg. For the initial positions of the pucks (i.e., x and y for each puck at time $t = 0.0$ s), compute the expected location of the center of mass using the formulas given in question #10. Check your answer using the simulation (be sure to run the simulation so that the new center of mass is computed).

18. For $m_2 = 0.6$ kg, compute the horizontal velocity of the center of mass using the formula, $u_{cm} = p_x/M$. Recall that you already computed p_x in question #7.

19. For $m_2 = 0.6$ kg, use your answers to questions #17 and #18, compute the expected location of the center of mass at a time $t = 2.0$ s using the formula,

$$x_{cm}(t) = x_{cm}(0) + u_{cm}t$$

Check your answer by measuring the location of the **X** mark.

Name: ______________________________ Instructor: ____________________

Partner: ______________________________ Section Day/Time: ______________

Project 15
Probability and Statistics

Objectives

In this project, you observe the motion of particles bouncing around in a box, a simple paradigm for gas atoms in a container. The topics covered in this project include:

- Probability
- Mean value and root-mean-square (r.m.s.) value
- Variance and standard deviation
- Histograms

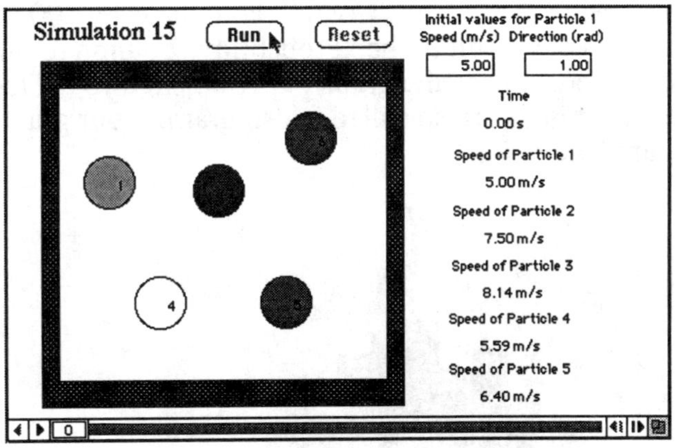

Procedure

Using the *Interactive Physics* program, run the simulation for this project. Five particles bounce around in a box; this system is representative of atoms in a gas.† As the particles move around, they collide with the walls and each other, altering their velocities. The motion of the particles serves to randomize their positions and velocities. The digital meters measure the speed of each particle and the elapsed time.

You should run the simulation and periodically (every 1.0 seconds) record the velocity of each particle. Since it would be very tedious to record the large quantity of data required to have a significant sample, you will record the velocities to only one significant digit.

In this project, we compute statistical quantities and answer questions like "what is the average speed of a particle?" and "what is the probability that a particle has a speed in a certain range?" The use of probability in physics is relatively new; only about 100 years ago did statistical calculations start to play an important role in physics.

---

† Atoms in air are smaller (diameters of about 10^{-9} m) and their speeds are higher (about 500 m/s at room temperature).

Questions and Problems

1. Set the speed of particle 1 to 5.00 m/s and its direction to 1.00 radians. Run the simulation and periodically record the velocity of each particle, filling out Table I. IMPORTANT: Record the particles' speeds only to a single significant digit, rounding to the nearest integer. For example, if the value is 6.34, record it as 6; if the value is 6.59, record it as 7.

2. Find the average value of the speed for the 100 values recorded in Table I. Compute this average by adding up the values in Table I and dividing by 100.

3. From your data in Table I, count the number of occurrences of each speed, that is, how many times did you record a particle having a given speed. Record the count in column 2 of Table II. When you have written down the count for each speed, sum up the counts and check that the sum equals 100 (which is the number of data entries in Table I).

4. In Table II, divide each value in the "Count" (column 2) by 100 (since you made 100 measurements) and record your result in the "Probability" column (column 3). Plot the probability versus speed; you can use the graph paper on page 104. Traditionally, this type of data is plotted using a bar graph (also called a histogram); your graph should look something like the figure below.

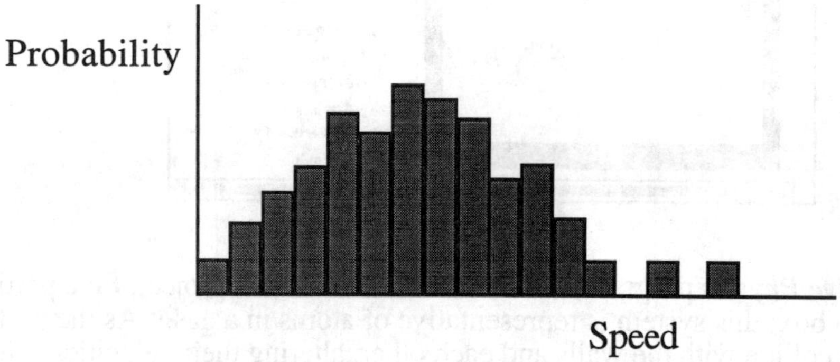

5. From your data in Table II, what was the most probable speed, v_{mp}? In other words, what speed had the highest count? Your histogram should have a peak at this value. If two or more bars are of equal height, take their average to estimate v_{mp}.

6. In Table II, multiply each entry in columns 1 and 3 together (i.e., "Speed" × "Probability") and record your result in column 4. After filling out column 4, compute the sum of the values in this column; record the tallied sum at the bottom of the column. This sum is $\overline{v}$, the average speed of the particles (also called the mean speed). Your value for $\overline{v}$ should match the value found in question #2; if it doesn't check your work.

Table I: Speeds of the particles at various times.

Time t (seconds)	Particle #1 (m/s)	Particle #2 (m/s)	Particle #3 (m/s)	Particle #4 (m/s)	Particle #5 (m/s)
1.0					
2.0					
3.0					
4.0					
5.0					
6.0					
7.0					
8.0					
9.0					
10.0					
11.0					
12.0					
13.0					
14.0					
15.0					
16.0					
17.0					
18.0					
19.0					
20.0					

Table II: Analysis of probability distribution.

1 (Data) Speed	2 (Data) Count	3 (Analysis) Probability	4 (Analysis) Speed × Prob.	5 (Analysis) $(\text{Speed})^2$	6 (Analysis) $(\text{Speed})^2$ × Prob.
0				0	
1				1	
2				4	
3				9	
4				16	
5				25	
6				36	
7				49	
8				64	
9				81	
10				100	
11				121	
12				144	
13				169	
14 (or more)				196	
Sum				----------	
Graph 1 Horiz. axis		Graph 1 Vertical axis	$\overline{v}$		$\overline{v^2}$

7. In Table II, multiply each entry in the columns 3 and 5 together (i.e., "Probability" × "(Speed)^2") and record your result in column 6. After filling out the column, compute the sum of the values in column 6; record the tallied sum at the bottom of the column. This sum is $\overline{v^2}$, the average square speed of the particles.

8. The root-means-square (r.m.s.) speed is defined as $\sqrt{\overline{v^2}}$, that is, as the square root of the average square speed. From your result of question #7, what is the r.m.s. speed of the particles?

9. The most probable speed, v_{mp}, the average speed, $\overline{v}$, and the r.m.s. speed, $\sqrt{\overline{v^2}}$, should be similar but not necessarily equal. What are their three values for your data?

10. The variance is defined as $\overline{v^2} - (\overline{v})^2$; the variance indicates the amount of variation that exists in the distribution of speeds (i.e., the "spread" in the histogram). Using your results from questions #6 and #7, compute the variance of the speed.

11. The standard deviation is defined as the square root of the variance, that is, $\sqrt{\overline{v^2} - (\overline{v})^2}$. The standard deviation also indicates the amount of variation (or spread) that occurs in the distribution of speeds. Using your results from question #10, compute the standard deviation of the speed.

12. Suppose the speed of every particles was *always* equal to 5. What would be the values of the most probable speed, v_{mp}, the average speed, $\overline{v}$, and the r.m.s. speed, $\sqrt{\overline{v^2}}$?

13. Again, suppose the speed of every particles was *always* equal to 5. What would be the values of the variance, $\overline{v^2} - (\bar{v})^2$, and the standard deviation, $\sqrt{\overline{v^2} - (\bar{v})^2}$?

14. Reset the simulation and change the initial direction of particle 1 to be 2.00 radians, keeping its initial speed at 5.00 m/s. Run the simulation and observe the motion. Without taking any data, which of the various quantities computed above (histogram, average speed, variance, etc.) would be significantly different. Explain your reasoning.

15. Reset the simulation and change the initial speed of particle 1 to be 15.0 m/s, with a direction angle of 1.00 radians. Run the simulation and observe the motion. Without taking any data, which of the various quantities computed from the original data set (histogram, average speed, variance, etc.) would be significantly different. Explain your reasoning.

Graph 1: Histogram of probability versus speed

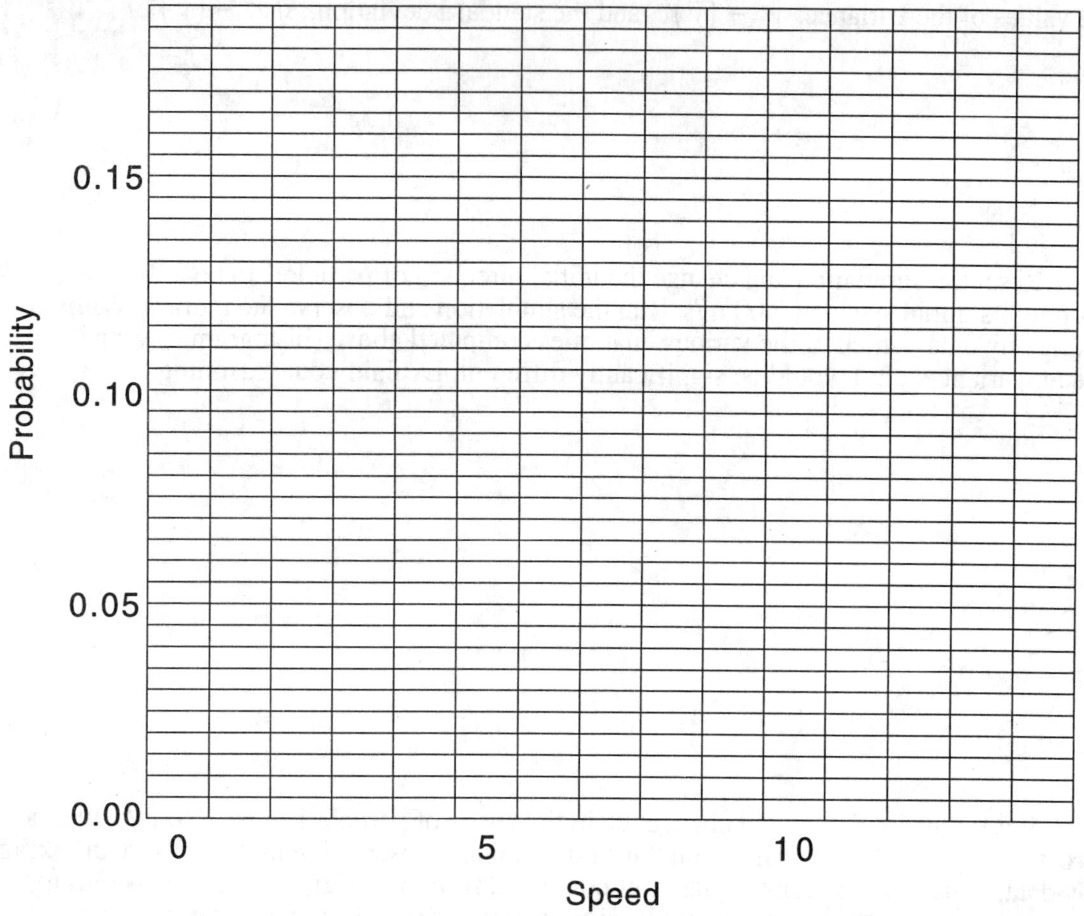